Vol. 68, No. 1 February 1995

MATHEMATICS MAGAZINE

EDITOR

Martha J. Siegel
   Towson State University

ASSOCIATE EDITORS

Donna Beers
   Simmons College

Douglas M. Campbell
   Brigham Young University

Paul J. Campbell
   Beloit College

Underwood Dudley
   DePauw University

Susanna Epp
   DePaul University

George Gilbert
   Texas Christian University

Judith V. Grabiner
   Pitzer College

David James
   Howard University

Dan Kalman
   American University

Loren C. Larson
   St. Olaf College

Thomas L. Moore
   Grinnell College

Bruce Reznick
   University of Illinois

Kenneth A. Ross
   University of Oregon

Doris Schattschneider
   Moravian College

Harry Waldman
   MAA, Washington, DC

EDITORIAL ASSISTANT

Dianne R. McCann

The *MATHEMATICS MAGAZINE* (ISSN 0025–570X) is published by the Mathematical Association of America at 1529 Eighteenth Street, N.W., Washington, D.C. 20036 and Montpelier, VT, bimonthly except July/August.

The annual subscription price for the MATHEMATICS MAGAZINE to an individual member of the Association is $16 included as part of the annual dues. (Annual dues for regular members, exclusive of annual subscription prices for MAA journals, are $64. Student and unemployed members receive a 66% dues discount; emeritus members receive a 50% discount; and new members receive a 40% dues discount for the first two years of membership.) The nonmember/library subscription price is $68 per year.

Subscription correspondence and notice of change of address should be sent to the Membership/Subscriptions Department, Mathematical Association of America, 1529 Eighteenth Street, N.W., Washington, D.C. 20036. Microfilmed issues may be obtained from University Microfilms International, Serials Bid Coordinator, 300 North Zeeb Road, Ann Arbor, MI 48106.

Advertising correspondence should be addressed to Ms. Elaine Pedreira, Advertising Manager, The Mathematical Association of America, 1529 Eighteenth Street, N.W., Washington, D.C. 20036.

Second class postage paid at Washington, D.C. and additional mailing offices.

Postmaster: Send address changes to Mathematics Magazine Membership/Subscriptions Department, Mathematical Association of America, 1529 Eighteenth Street, N.W., Washington, D.C. 20036-1385.

PRINTED IN THE UNITED STATES OF AMERICA

ARTICLES

The Roots of Commutative Algebra in Algebraic Number Theory

ISRAEL KLEINER
York University
North York, Ontario, Canada M3J 1P3

Introduction

The concepts of field, (commutative) ring, ideal, and unique factorization are among the fundamental notions of (commutative) algebra. How did they arise? In large measure, from three central problems in number theory: Fermat's Last Theorem, reciprocity laws, and binary quadratic forms. In this paper we will describe how this happened.

To put the issues in a broader context, these three number-theoretic problems were instrumental in the emergence of algebraic number theory—one of the two main sources of the modern discipline of commutative algebra.[1] The other source was algebraic geometry. It was in the setting of these two subjects that many of the main concepts and results of commutative algebra evolved in the 19th century. Thus commutative algebra can be said to have been well developed before it was created.

To set the scene for our story, a word about mathematics, and especially algebra, in the 19th century. The period witnessed fundamental transformations in mathematics —in its concepts, its methods, and in mathematicians' attitude toward their subject. These included a growing insistence on rigor and abstraction; a predisposition for founding general theories rather than focusing on specific problems; an acceptance of nonconstructive existence proofs; the emergence of mathematical specialties and specialists; the rise of the view of mathematics as a free human activity, neither deriving from, nor dependent on, nor necessarily applicable to concrete setting; the rise of set-theoretic thinking; and the reemergence of (a new variant of) the axiomatic method. As for algebra, its subject-matter and methods changed beyond recognition. Algebra as the study of solvability of polynomial equations gave way to algebra as the study of abstract structures defined axiomatically. While in the past algebra was on the periphery of mathematics, it now (the 19th and early 20th centuries) became one of its central concerns. Moreover, algebra began to penetrate other mathematical fields (e.g. geometry, analysis, logic, topology, number theory) to such an extent that in the early decades of the 20th century one began to speak of the algebraization of mathematics ([3, p. 135] and [22]).

---

[1]Commutative algebra is nowadays understood to be the study of commutative noetherian rings. The first book on the subject, dating from the 1950s, was Zariski and Samuel's *Commutative Algebra*. In the first decades of the 20th century, what we understand as commutative algebra was known as Ideal Theory, a name which reflects the sources of the subject.

Algebraic Number Theory and Unique Factorization

Algebraic number theory arose (as noted) from the investigation of three fundamental problems: Fermat's Last Theorem, reciprocity laws, and binary quadratic forms. Although these problems had their roots in the 17th and 18th centuries, they began to be intensively studied only in the 19th century. The strategy that began to emerge was to embed the domain Z of integers, in terms of which these problems were formulated, in domains of (what came to be known as) algebraic integers. Let me explain.

Fermat's Last Theorem states that $x^n + y^n = z^n$ has no nontrivial integer solutions for $n > 2$. It suffices to consider $n = 4$ and $n = p$, an odd prime. Fermat proved the conjecture for $n = 4$ in the early 17th century, and Euler proved it for $n = 3$ in the 18th (Euler's proof had a gap). In the early 19th century, Dirichlet and Legendre proved Fermat's conjecture for $n = 5$, but the case $n = 7$ defied the efforts of the best mathematicians, including Dirichlet and Gauss.[2] A new approach was called for. It was provided by Lamé.

On March 1, 1847, Lamé excitedly announced before the Paris Academy of Sciences that he had proved Fermat's Last Theorem. His basic idea was to factor the left-hand side of $x^p + y^p = z^p$ as follows:

$$x^p + y^p = (x + y)(x + \omega y)(x + \omega^2 y) \cdots (x + \omega^{p-1} y), \qquad (1)$$

where ω is a primitive pth root of 1. Since $x^p + y^p = z^p$, the product of the factors on the right-hand side of (1) is a pth power. Lamé claimed that if these factors are relatively prime, then each is a pth power. From this a contradiction can be derived using Fermat's method of infinite descent. If the factors $x + y, x + \omega y, \ldots, x + \omega^{p-1} y$ are not relatively prime, then, upon division by a suitable a, we obtain the relatively prime factors

$$\frac{x + y}{a}, \frac{x + \omega y}{a}, \ldots, \frac{x + \omega^{p-1} y}{a},$$

and the proof, Lamé asserted, proceeds analogously.[3]

Following Lamé's presentation, Liouville, who was in the audience, took the floor and pointed to what seemed to him to be a gap in Lamé's proof. The gap, he said, was Lamé's contention that if a product of relatively prime factors is a pth power, then each of the factors must be a pth power. This result is of course true for the integers, Liouville noted, but it required a proof for the complex entities Lamé was dealing with.

Liouville's observation went to the heart of what was wrong with Lamé's proof. Specifically, to prove Lamé's claim one had to show that unique factorization holds in the domain

$$D_p = \left\{ a_0 + a_1 \omega + \cdots + a_{p-1} \omega^{p-1} : a_i \in Z \right\}$$

of so-called cyclotomic integers. About three months after Lamé's presentation

---

[2] Gauss was apparently not very interested in Fermat's problem. In a letter to Olbers in 1816 he said: "I confess indeed that Fermat's theorem as an isolated proposition has little interest for me, since a multitude of such propositions, which one can neither prove nor refute, can easily be formulated" [**20**, pp. 818–819].

[3] The essential idea of Lamé's approach is contained in the following method for finding all primitive pythagorean triples (x, y, z): Factoring the left-hand side of $x^2 + y^2 = z^2$ we get $(x + yi)(x - yi) = z^2$. It can be shown that $x + yi$ and $x - yi$ are relatively prime (since $(x, y, z) = 1$), and since their product is a square, each is a square. In particular, $x + yi = (a + bi)^2 = (a^2 - b^2) + 2abi$, hence $x = a^2 - b^2$, $y = 2ab$, and it follows that $z = a^2 + b^2$. This is the standard formula yielding all pythagorean triples.

Kummer wrote to Liouville that D_p is in general *not* a unique factorization domain (UFD). (The first p for which it is not is $p = 23$.)

It is important to note that although Lamé's proof was false (it *is* correct for those p for which D_p is a UFD), his overall strategy proved fundamental for subsequent approaches to Fermat's Last Theorem. In particular, he proved the theorem for $n = 7(!)$, a feat that (we recall) eluded the likes of Dirichlet and Gauss.

We turn now to reciprocity laws.[4] Gauss was the first to define formally the notion of congruence. Thus it was natural for him to pose the question of solvability of the congruence $a_0 + a_1 x + a_2 x^2 + \cdots + a_m x^m \equiv 0 \pmod{n}$. The problem in this generality proved intractable (it is so to this day), and so (as Pólya would advise) Gauss considered a simpler case, namely $a_0 + a_1 x + a_2 x^2 \equiv 0 \pmod{n}$. Several reductions show that it suffices to consider the congruence $x^2 \equiv q \pmod{p}$, where both p and q are odd primes (the case of even primes has to be considered separately). Gauss proved that there is a "reciprocity relation" between the solvability of $x^2 \equiv q \pmod{p}$ and $x^2 \equiv p \pmod{q}$. Specifically, $x^2 \equiv q \pmod{p}$ is solvable if, and only if, $x^2 \equiv p \pmod{q}$ is solvable, unless $p \equiv q \equiv 3 \pmod{4}$. In the latter case, $x^2 \equiv q \pmod{p}$ is solvable if, and only if, $x^2 \equiv p \pmod{q}$ is not. This is the celebrated quadratic reciprocity law, one of the "jewels of number theory," proved in Gauss' *Disquisitiones Arithmeticae* of 1801.

What about higher reciprocity laws? That is, is there a "reciprocity relation" between the solvability of $x^m \equiv q \pmod{p}$ and $x^m \equiv p \pmod{q}$ for $m > 2$? Gauss opined that such laws cannot even be properly conjectured within the context of natural numbers [28, p. 105], adding that

> The previously accepted laws of arithmetic are not sufficient for the foundations of a general theory ... Such a theory demands that the domain of higher arithmetic be endlessly enlarged [17, p. 108].

A prophetic statement, indeed. Gauss was calling (in modern terms) for the founding of an arithmetic theory of algebraic numbers. In fact, Gauss himself began to enlarge the domain of arithmetic by introducing what came to be known as the Gaussian integers,

$$Z[i] = \{a + bi : a, b \in Z\},$$

and showing that they form a UFD. This he did in two papers in 1829 and 1831, in which he used $Z[i]$ to formulate the law of biquadratic reciprocity.[5] In the 1831

---

[4]It is no easy matter to define what a reciprocity law is (see [10], [21], [29]). For our purposes, the descriptions that follow suffice.

[5]The law can be stated (essentially) as follows [8, p. 64]:

$$\left(\frac{\pi}{\sigma}\right)_4 \left(\frac{\sigma}{\pi}\right)_4 = (-1)^{[(N(\pi)-1)/4][(N(\sigma)-1)/4]},$$

where π and σ are distinct primes in $Z[i]$, $N(\alpha)$ is the norm of the Gaussian integer α, and

$$\left(\frac{\alpha}{\pi}\right)_4 = \begin{cases} 1 \text{ if } x^4 \equiv \alpha \pmod{\pi} \text{ is solvable in } Z[i] \\ -1, i, \text{ or } -i \text{ otherwise} \end{cases}$$

(We recall that the quadratic reciprocity law can be stated in terms of the Legendre symbol as

$$\left(\frac{p}{q}\right)\left(\frac{q}{p}\right) = (-1)^{[(p-1/2)][(q-1)/2]}.)$$

The *proof* of the biquadratic reciprocity law was found in Gauss' diaries. The first published proof was given by Eisenstein in 1844. The theorem can be used to determine the solvability in Z of $x^4 \equiv q \pmod{p}$ [17, p. 127].

paper Gauss also introduced the geometric representation of complex numbers as points in the Euclidean plane. At about the same time, Jacobi and Eisenstein (as well as Gauss in unpublished papers) formulated the cubic reciprocity law. Here one needed to consider the domain

$$Z[\rho] = \{a + b\rho : a, b \in Z\},$$

where ρ is a primitive cube root of 1. $Z[\rho]$ was also shown to be a UFD. The search was on for higher reciprocity laws. But as in the case of Fermat's Last Theorem, here too one needed new methods to deal with cases beyond the first few, since unlike $Z[i]$ and $Z[\rho]$, other domains of higher arithmetic needed to formulate such laws were not UFDs.

We now turn to the third number-theoretic problem related to the rise of algebraic number theory, namely representation of integers by binary quadratic forms. (This problem is, in fact, intimately connected to reciprocity laws; see [10].) An (integral) binary quadratic form is an expression of the form $f(x, y) = ax^2 + bxy + cy^2$, $a, b, c \in Z$. The major problem of the theory of quadratic forms was: Given a form f, find all integers m that can be represented by f; that is, for which $f(x, y) = m$. This problem was studied for specific f by Fermat, and intensively investigated by Euler. For example, Fermat considered the representation of integers as sums of two squares. It was, however, Gauss in the *Disquisitiones* who made the fundamental breakthroughs and developed a comprehensive and beautiful theory of binary quadratic forms. Most important was his definition of the *composition* of two forms and his proof that the (equivalence classes of) forms with a given "discriminant" $D = b^2 - 4ac$ form a commutative group under this composition.

The *idea* behind composition of forms is simple: If forms f and g represent integers m and n, respectively, then their composition $f * g$ should represent the product mn. The *implementation* of this idea is subtle and "extremely difficult to describe" [12, p. 334]. Attempts to gain conceptual insight into Gauss' theory of composition of forms inspired the efforts of some of the best mathematicians of the time, among them Dirichlet, Kummer, and Dedekind. The key idea here, too, was to extend the domain of higher arithmetic and view the problem in a broader context. Here is perhaps the simplest illustration: If m_1 and m_2 are sums of two squares, so is $m_1 m_2$. Indeed, if $m_1 = x_1^2 + y_1^2$ and $m_2 = x_2^2 + y_2^2$, then $m_1 m_2 = (x_1 x_2 - y_1 y_2)^2 + (x_1 y_2 + x_2 y_1)^2$. In terms of the composition of quadratic forms this can be expressed as $f(x_1, y_1) * f(x_2, y_2) = f(x_1 x_2 - y_1 y_2, x_1 y_2 + x_2 y_1)$, or $f * f = f$, where $f(x, y) = x^2 + y^2$. But even this "simple" law of composition seems mysterious and ad hoc until one introduces Gaussian integers, which make it transparent ($\bar{\alpha}$ denotes the conjugate of α):[6]

$$
\begin{aligned}
(x_1^2 + y_1^2)(x_2^2 + y_2^2) \\
&= (x_1 + y_1 i)(x_1 - y_1 i)(x_2 + y_2 i)(x_2 - y_2 i) \\
&= (x_1 + y_1 i)(x_2 + y_2 i)\overline{(x_1 + y_1 i)(x_2 + y_2 i)} \\
&= [(x_1 x_2 - y_1 y_2) + (x_1 y_2 + x_2 y_1)i]\overline{[(x_1 x_2 - y_1 y_2) + (x_1 y_2 + x_2 y_1)i]} \\
&= (x_1 x_2 - y_1 y_2)^2 + (x_1 y_2 + x_2 y_1)^2.
\end{aligned}
$$

In general, $ax^2 + bxy + cy^2 = m$ can be written as

$$\frac{1}{a}\left(ax + \frac{b + \sqrt{D}}{2} y\right)\left(ax + \frac{b - \sqrt{D}}{2} y\right) = m,$$

---

[6]The following example gives a better sense of the subtlety and apparent arbitrariness of Gauss' law of composition of binary quadratic forms: If $g(x, y) = 2x^2 + 2xy + 3y^2$, then $g(x_1, y_1) * g(x_2, y_2) = h(2x_1 x_2 + x_1 y_2 + x_2 y_1 - 2y_1 y_2, x_1 y_2 + x_2 y_1 + y_1 y_2)$, where $h(x, y) = x^2 + 5y^2$, so $g * g = h$.

where $D = b^2 - 4ac$ is the discriminant of the quadratic form. We have thus expressed the problem of representation of integers by binary quadratic forms in terms of the domain

$$R = \left\{ \frac{u + v\sqrt{D}}{2} : u, v \in Z, u \equiv v \pmod{2} \right\}.$$

Specifically, if we let

$$a = \alpha, \quad \frac{b + \sqrt{D}}{2} = \beta,$$

then

$$ax^2 + bxy + cy^2 = \frac{1}{a}(\alpha x + \beta y)(\bar{\alpha}x + \bar{\beta}y) = \frac{1}{a}N(\alpha x + \beta y),$$

where N denotes the *norm*. Thus to solve $ax^2 + bxy + cy^2 = m$ is to find $x, y \in Z$ such that $N(\alpha x + \beta y) = m$. In fact, Kummer noted in the 1840s that the entire theory of binary quadratic forms can be regarded as the theory of "complex numbers" (Kummer's terminology) of the form $x + y\sqrt{D}$ [6, p. 585]. And since such domains of "complex numbers" did not, in general, possess unique factorization, the development of their arithmetic theory became an important goal. (Gauss' theory of binary quadratic forms may, in fact, be seen as the first major (implicit) attempt to deal with nonunique factorization.)

It is time to take stock. We have seen that in dealing with central problems in number theory, namely Fermat's Last Theorem, reciprocity laws, and binary quadratic forms, it was found important to formulate them as problems in domains of (what came to be called) algebraic integers. This often transformed additive problems into multiplicative ones (e.g. $x^3 + y^3 = z^3$ was changed to $(x + y)(x + \rho y)(x + \rho^2 y) = z^3$, ρ a primitive cube root of unity). Now, multiplicative problems in number theory are, in general, much easier to handle than additive problems—provided that one has the (multiplicative) machinery to do it, namely unique factorization in the domain under consideration. Thus the study of unique factorization in domains of algebraic integers became the major problem of a newly emerging subject—algebraic number theory. Number theory had lost its innocence (which, at the appropriate time, is of course no bad thing). After more than 2000 years in which number theory meant the study of properties of the (positive) integers, its scope became enormously enlarged. One could no longer use the term "integer" with impunity. The term had to be qualified —a rational (ordinary) integer, a Gaussian integer, a cyclotomic integer, a quadratic integer, or any one of an infinite species of other algebraic integers. Of course the real difficulty was not the multiplicity of types of integers, but the fact that, in general, domains of such integers no longer possessed unique factorization.

Could one regain the paradise lost by establishing unique factorization in *some* sense for these domains? An affirmative answer was given by Dedekind and Kronecker in the last third of the 19th century. We will describe the work of Dedekind, who introduced the notions of ideal and prime ideal, and showed that every nonzero ideal in an arbitrary domain of algebraic integers is a unique product of prime ideals. (An ideal P is prime if $xy \in P$ implies $x \in P$ or $y \in P$.) This result generalizes the theorems on unique factorization of elements into primes in (for example) Z, Z[i], and Z[ρ]. But it was preceded by very significant contributions by Kummer to the problem of unique factorization, which inspired both Dedekind and Kronecker (in different ways). We begin with a sketch of Kummer's contribution.

Kummer and Ideal Numbers

We recall that the domains of cyclotomic integers $D_p = \{a_0 + a_1\omega + \cdots + a_{p-1}\omega^{p-1}$: $a_i \in Z$, ω a primitive pth root of $1\}$ were central in the study of Fermat's conjecture. They also proved important in the investigation of higher reciprocity laws. The two problems are, in fact, related, and progress on one was often linked with progress on the other (see [17, p. 203]). Gauss contended that if he were able to derive higher reciprocity laws, then Fermat's Last Theorem would be one of the less interesting corollaries. That is one prediction where Gauss proved to be wrong: A general reciprocity law (due to Artin) was obtained in the 1920s, but Fermat's problem is not resolved.[7]

Both Fermat's problem and higher reciprocity laws were of great interest to Kummer (the latter apparently more than the former), and to make serious progress on them it was essential to establish unique factorization (of some type) in the domains D_p. This Kummer accomplished in the 1840s. As he put it in a letter to Liouville, unique factorization in D_p "can be saved by the introduction of a new kind of complex numbers that I have called ideal complex numbers" [3, p. 101]. Kummer then showed that every element in the domain of cyclotomic integers is a unique product of "ideal primes".

Kummer's theory was vague and computational. In fact, the central notion of ideal number was only *implicitly* defined in terms of its divisibility properties (see [12] or [13]).[8] Kummer noted that in adopting the implicit definition he was guided by the idea of "free radical" in chemistry, a substance whose existence can only be discerned by its effects [9, p. 35].

The following example (even if not of a cyclotomic domain) may give a sense of Kummer's theory of ideal numbers. Let $D = \{a + b\sqrt{5}\,i: a, b \in Z\}$. This is a standard example (due to Dedekind) of a domain in which factorization is not unique. For example, $6 = 2 \times 3 = (1 + \sqrt{5}\,i)(1 - \sqrt{5}\,i)$, where $2, 3$, and $1 \pm \sqrt{5}\,i$ are primes (indecomposables) in D [1, p. 250]. To restore unique factorization of $6 \in D$, adjoin the "ideal numbers" $\sqrt{2}$, $(1 + \sqrt{5}\,i)/\sqrt{2}$, and $(1 - \sqrt{5}\,i)/\sqrt{2}$. These are, in fact, *ideal primes*.[9] We then have

$$6 = 2 \times 3 = \sqrt{2} \times \sqrt{2} \times \frac{1 + \sqrt{5}\,i}{\sqrt{2}} \times \frac{1 - \sqrt{5}\,i}{\sqrt{2}},$$

---

[7]These unprophetic words were of course uttered prior to Wiles' June 1993 announcement of his proof of Fermat's Last Theorem. In a widely distributed e-mail note in December 1993 Wiles noted that "during the review process a number of problems emerged" but that he believes that he "will be able to finish [closing the gaps in the proof] in the near future." In any case, Wiles' proof (when (if) the gaps are closed) would by no means follow from Artin's reciprocity law.

[8]Every ideal number of D_p is of the form $\sqrt[h]{\alpha}$, where $\alpha \in D_p$ [3, p. 105], although not every number of this form is an ideal number. The positive integer h (which is independent of α) is called the *class number* of D_p. It is a measure of "how far" D_p is removed from possessing unique factorization: $h = 1$ if, and only, if D_p is a UFD. In modern terms, the ideal numbers (of D_p) are in one-one correspondence with the valuations on the cyclotomic field $Q_p = \{q_0 + q_1\omega + \cdots + q_{p-1}\omega^{p-1}: q_i \in Q\}$ (see [6, p. 581]).

[9]An ideal prime had no more explicit definition than an ideal number. It was described in terms of the set of all cyclotomic integers divisible by it [13, p. 325]. Note also that the adjoined ideal numbers are square roots of elements of D, since

$$\frac{1 \pm \sqrt{5}\,i}{\sqrt{2}} = \sqrt{-2 \pm \sqrt{5}\,i}\,,$$

hence D is "not too far removed" from being a unique factorization domain (assuming that this is true for all elements of D); that is, its "class number" is 2.

and $6 = (1 + \sqrt{5}i)(1 - \sqrt{5}i) = \sqrt{2} \times (1 + \sqrt{5}i)/\sqrt{2} \times \sqrt{2} \times (1 - \sqrt{5}i)/\sqrt{2}$; that is, the decomposition of 6 into primes is now unique. Furthermore, although the choice of the ideal primes $\sqrt{2}$, $(1 \pm \sqrt{5}i)/\sqrt{2}$ seems to have been ad hoc, it will come to seem natural after ideals are introduced.

A cruder but perhaps more revealing example is the "domain" $H = \{3k + 1:$ $k = 0, 1, 2, \ldots\} = \{1, 4, 7, 10, 13, \ldots\}$, introduced by Hilbert. (This is not an integral domain, but its structure as a multiplicative semigroup illustrates the ideas involved.) Here $100 = 10 \times 10 = 4 \times 25$, where 4, 10, and 25 are primes of H. We can eliminate the nonuniqueness of this factorization by adjoining the "ideal primes" 2 and 5. Then $100 = 10 \times 10 = 2 \times 5 \times 2 \times 5$ and $100 = 4 \times 25 = 2 \times 2 \times 5 \times 5$.

The notion of adjoining elements to a given mathematical structure in order to obtain a desired property is commonplace and important in various areas of mathematics. Thus we adjoin $\sqrt{-1}$ to the reals to obtain algebraic closure, and adjoin "ideal points" (points at infinity) to the euclidean plane to obtain symmetry (duality) between points and lines. The adjunction of ideal numbers to the cyclotomic integers should be seen in the same light.

Kummer's ideas were brilliant but difficult and not clearly formulated. The fundamental concepts of ideal number and ideal prime were not intrinsically defined. "Today," notes Edwards, "few mathematicians would find [Kummer's] definition [of ideal primes] acceptable" [15, p. 6]. Most importantly, Kummer's decomposition theory was devised only for cyclotomic integers (he mentions, without elaboration, that an analogous theory could be developed for algebraic integers of the form $x + y\sqrt{D}$). What was needed was a decomposition theory that would apply to arbitrary domains of algebraic integers. This called for a fundamentally new approach to the subject, provided (independently, and in different ways) by Dedekind and Kronecker. We will focus (as mentioned) on Dedekind's formulation, which is the one that has generally prevailed.

Dedekind and Ideals

Dedekind's work (in 1871)[10] was revolutionary in several ways: in its formulation, its grand conception, its fundamental new ideas, and its modern spirit. Its main result was that every nonzero ideal in the domain of integers of an algebraic number field is a unique product of prime ideals. Before one could state this theorem one had, of course, to define the concepts in its statement, especially "the domain of integers of an algebraic number field" and "ideal." It took Dedekind about 20 years to formulate them.

The number-theoretic domains studied at the time, such as the Gaussian integers, the integers arising from cubic reciprocity, and the cyclotomic integers, are all of the form $Z[\theta] = \{a_0 + a_1\theta + \cdots + a_n\theta^n: a_i \in Z\}$, where θ satisfies an "appropriate" polynomial with integer coefficients. It was therefore tempting to define the domains to which Dedekind's theorem would apply as objects of this type. These were, however, the wrong objects, Dedekind showed. For example, he proved that Kummer's theory of unique factorization could *not* be extended to the domain $Z[\sqrt{3}\,i] = \{a + b\sqrt{3}\,i: a, b \in Z\}$ [13, p. 337], and, of course, Dedekind's objective was to try to extend Kummer's theory to *all* domains of algebraic integers. One had to begin the search for the appropriate domains, Dedekind contended, within an "algebraic

[10] It appeared as Supplement X to the 2nd edition of Dirichlet's *Vorlesungen über Zahlentheorie*, which was written as commentary on, and explication of, Gauss' *Disquisitiones*. Dedekind wrote two other versions of his theory, in Supplements to the 3rd and 4th editions of Dirichlet's *Zahlentheorie* (in 1879 and 1894 respectively).

number field"—a finite field extension $Q(\alpha) = \{q_0 + q_1\alpha + \cdots + q_s\alpha^s : q_i \in Q\}$ of the rationals, α an algebraic number (the notion of "algebraic number" was well known at the time, but not that of "algebraic integer"). He showed that $Q(\alpha)$ is closed under the four algebraic operations of addition, subtraction, multiplication, and division, and then defined axiomatically (for the first time) the notion of a *field*:[11]

A system A of real or complex numbers is called a field if the sum, difference, product, and quotient of any two numbers of A belongs to A [3, p. 118].

This procedure was typical of Dedekind's modus operandi: He would distill from a concrete object (in this case $Q(\alpha)$) the properties of interest to him (in this case closure under the four algebraic operations) and proceed to define an abstract object (in this case a field) in terms of those properties. This is, of course, standard practice nowadays, but it was revolutionary in Dedekind's time. Dedekind used it again and again, as we shall see.

To come back to the domain of algebraic integers to which Dedekind's result was to apply: He defined it to be the set R of all elements of $Q(\alpha)$ that are roots of *monic* polynomials with integer coefficients. (*All* the elements of $Q(\alpha)$ are roots of polynomials, not necessarily monic, with integer coefficients.) He showed that these elements "behave" like integers—they are closed under addition, subtraction, and multiplication. They are the "integers" of the algebraic number field $Q(\alpha)$. Dedekind did not, however, motivate this basic definition of the domain of algebraic integers, a fact lamented by Edwards: "Insofar as this is the crucial idea of the theory, the genesis of the theory appears, therefore, to be lost" [13, p. 332].

This is not to say, of course, that the notion of the domain of algebraic integers cannot be motivated. The (algebraic) integers of $Q(\alpha)$ are to $Q(\alpha)$ what the (rational) integers (the elements of Z) are to Q. Now, the elements of Q can be thought of as roots of *linear* polynomials with integer coefficients (viz. $p(x) = ax + b$). Among these, the integers are the roots of *monic* linear polynomials (viz. $q(x) = x + b$). If we extend this analysis to polynomials of arbitrary degree (over Z), their roots (in $Q(\alpha)$) yield (all of) $Q(\alpha)$, while the roots of the monic polynomials among them yield the domain of integers of $Q(\alpha)$. Moreover, just as Q is the field of quotients of Z, so $Q(\alpha)$ is the field of quotients of *its* domain of integers. (See also [27, pp. 264–265], for a discussion, from a different perspective, of the issue of motivation of algebraic integers.)

Having defined the domain R of algebraic integers (of $Q(\alpha)$) in which he would formulate and prove his result on unique decomposition of ideals, Dedekind considered, more generally, sets of integers of $Q(\alpha)$ closed under addition, subtraction, and multiplication. He called them "orders". (The domain R of integers of $Q(\alpha)$ is the largest order.) Here, then, was another algebraic first for Dedekind—an essentially axiomatic definition of a (commutative) ring, albeit in a concrete setting.[12]

The second fundamental concept of Dedekind's theory, that of ideal, derived its motivation (and name) from Kummer's ideal numbers. Dedekind wanted to characterize them *internally*, within the domain D_p of cyclotomic integers. Thus, for each ideal number σ he considered the set of cyclotomic integers divisible by σ. These, he noted, are closed under addition and subtraction, as well as under multiplication by all elements of D_p. Conversely, he proved (and this is a difficult theorem) that every

---

[11]Algebraic number theory was not the only source of the field concept. Others were Galois theory and algebraic geometry. See [24].

[12]Algebraic number theory was *one* of the sources of commutative ring theory. The other was algebraic geometry. Noncommutative ring theory had entirely different origins. See [19].

set of cyclotomic integers closed under these operations is precisely the set of cyclotomic integers divisible by some ideal number τ. Thus there is a one-one correspondence between ideal numbers and subsets of the cyclotomic integers closed under the above operations. Such subsets of D_p Dedekind called *ideals*. These subsets, then, characterized ideal numbers internally, and served as motivation for the introduction of ideals in arbitrary domains of algebraic integers.[13] Dedekind defined them abstractly as follows [**13**, p. 343]:

A subset I of the integers R of an algebraic number field K is an *ideal* of R if it has the following two properties:
  (i) If $\beta, \gamma \in I$ then $\beta \pm \gamma \in I$.
  (ii) If $\beta \in I$, $\mu \in R$ then $\beta \mu \in I$.

Dedekind then defined a prime ideal—perhaps the most important notion of commutative algebra—as follows: An ideal P of R is *prime* if its only divisors are R and P. Given ideals A and B, A was said to divide B if $A \supseteq B$. In later versions of his work Dedekind showed that A divides B if, and only if, $B = AC$ for some ideal C of R (see [**7**, p. 122]). Having defined the notion of prime ideal, Dedekind proved his fundamental theorem that every nonzero ideal in the ring of integers of an algebraic number field is a unique product of prime ideals.

Here is another way of introducing ideals, in the spirit of *Kronecker's* work on establishing unique factorization in domains of algebraic integers. It is perhaps best illustrated with an example. Let $D = \{a + b\sqrt{5}\,i \colon a, b \in Z\}$. As we noted, D is not a UFD. Since uniqueness of factorization in a domain is equivalent to the existence of the greatest common divisor (g.c.d.) of any two nonzero elements in the domain [**16**, p. 199], it follows that one way to reestablish unique factorization in D is to introduce g.c.d.s in D. (This is Kronecker's motivation for the introduction of divisors (ideals); see [**5**, p. 125], or [**13**, p. 353].) How do we introduce the g.c.d. of, say, 2 and $1 + \sqrt{5}\,i$, which appear in the factorizations of 6 into primes, $6 = 2 \times 3 = (1 + \sqrt{5}\,i)(1 - \sqrt{5}\,i)$? We look at Z (or any UFD) for guidance.

If $a, b \in Z$, the g.c.d. of a and b can be found among the elements of the set $I = \{ax + by \colon x, y \in Z\}$. In fact, I is a principal ideal of Z, $I = \langle d \rangle = \{dm \colon m \in Z\}$, and g.c.d. $(a, b) = d$. (An ideal J of a ring R is principal if it is generated by a single element $a \in R$. We write $J = \langle a \rangle$.) In like manner, if D were a UFD, then g.c.d. $(2, 1 + \sqrt{5}\,i)$ would be found among the elements of $P = \{2\alpha + (1 + \sqrt{5}\,i)\beta \colon \alpha, \beta \in D\}$. Moreover, we would have $P = \langle s \rangle$ (the principal ideal of D generated by some $s \in D$) and g.c.d. $(2, 1 + \sqrt{5}\,i) = s$. Since D is not a UFD, there is no $s \in D$ such that $P = \langle s \rangle$. But why not have all of P represent (capture) g.c.d. $(2, 1 + \sqrt{5}\,i)$? Indeed, this is Kronecker's idea. (Dedekind would have argued for an "ideal element" $\hat{s}$ ($\hat{s} \notin D$) to describe P, so that P would be the set of elements in D divisible by $\hat{s}$; we exhibit such an $\hat{s}$ below.) We note that P is an ideal of D, though not a principal ideal. Similarly we let $Q = \langle 3, 1 + \sqrt{5}\,i \rangle = \{3\alpha + (1 + \sqrt{5}\,i)\beta \colon \alpha, \beta \in D\}$ and $R = \langle 3, 1 - \sqrt{5}\,i \rangle$ (note that $\langle 2, 1 - \sqrt{5}\,i \rangle = \langle 2, 1 + \sqrt{5}\,i \rangle$). We then easily verify that $P^2 = \langle 2 \rangle$, $PQ = \langle 1 + \sqrt{5}\,i \rangle$, $QR = \langle 3 \rangle$, and $PR = \langle 1 - \sqrt{5}\,i \rangle$. (If A and B are ideals of a ring R, their product is the ideal $AB = \{\sum_{\text{finite}} a_i b_i \colon a_i \in A, \, b_i \in B\}$.)

Returning now to the factorizations $6 = 2 \times 3 = (1 + \sqrt{5}\,i)(1 - \sqrt{5}\,i)$, they yield the following factorizations of ideals: $\langle 6 \rangle = \langle 2 \rangle \langle 3 \rangle = P^2(QR)$ and $\langle 6 \rangle = \langle 1 + \sqrt{5}\,i \rangle \langle 1 - \sqrt{5}\,i \rangle = (PQ)(PR) = P^2 QR$. One can readily verify that the ideals P, Q, R are prime. Thus the *ideal* $\langle 6 \rangle$ (if not the *element* 6) has been factored uniquely into prime ideals. Paradise regained. See [**5**], [**13**].

---

[13] Ideals also arose in an entirely different setting in noncommutative algebra. They appeared in the works of Cartan (1898) and especially of Wedderburn (1907), who called them "invariant subalgebras". See [**19**].

Let us compare this factorization of $\langle 6 \rangle$ into *prime ideals* with the factorization of 6 into *ideal primes* (à la Kummer) that we gave earlier:

$$6 = 2 \times 3 = \sqrt{2} \times \sqrt{2} \times \frac{1 + \sqrt{5}\,i}{\sqrt{2}} \times \frac{1 - \sqrt{5}\,i}{\sqrt{2}},$$

and

$$6 = \left(1 + \sqrt{5}\,i\right)\left(1 - \sqrt{5}\,i\right) = \sqrt{2} \times \frac{1 + \sqrt{5}\,i}{\sqrt{2}} \times \sqrt{2} \times \frac{1 - \sqrt{5}\,i}{\sqrt{2}}.$$

Performing some 18th-century symbolic callisthenics, we obtain the following: Since $P^2 = \langle 2 \rangle$, $P \sim \sqrt{2}$ (where " $\sim$ " stands for "corresponds to", "captures," "represents"). In fact, P is the ideal consisting of all elements of D divisible by the ideal number $\sqrt{2}$ (that is, such that the quotient is an algebraic integer [7, p. 235]). This would have been Dedekind's way of introducing P (that is, $\hat{s} = \sqrt{2}$). We also have $PQ = \langle 1 + \sqrt{5}\,i \rangle$, hence $PQ/P \sim (1 + \sqrt{5}\,i)/\sqrt{2}$, so that the ideal Q corresponds to the ideal number $(1 + \sqrt{5}\,i)/\sqrt{2}$. And since $PR = \langle 1 - \sqrt{5}\,i \rangle$, $PR/R \sim (1 - \sqrt{5}\,i)/\sqrt{2}$, hence $R \sim (1 - \sqrt{5}\,i)/\sqrt{2}$. This removes the mystery associated with our earlier introduction of the ideal numbers $\sqrt{2}$ and $(1 \pm \sqrt{5}\,i)/\sqrt{2}$. (How might Dedekind have restored unique factorization to the "domain" $H = \{1, 4, 7, 10, 13, \ldots\}$ we considered earlier?)

Dedekind's Legacy

Dedekind's Supplement X to Dirichlet's *Zahlentheorie* was the culmination of 70 years of investigations of problems related to unique factorization. It created, in one swoop, a new subject—algebraic number theory. It introduced, albeit in a concrete setting, some of the most fundamental concepts of commutative algebra, such as field, ring, ideal, prime ideal, and module (the last we have not discussed). These became basic in algebra and beyond. Supplement X also established one of the central results of algebraic number theory, namely the representation of ideals in domains of integers of algebraic number fields as unique products of prime ideals. The theorem was soon to play a fundamental role in the study of algebraic curves. It would also serve as a model for decomposition results in algebra.

As important as his concepts and results were Dedekind's methods. In fact, "his insistence on philosophical principles was responsible for many of his important innovations" [13, p. 349]. One of his philosophical principles was a focus on intrinsic, conceptual properties over formulas, calculations, or concrete representations. Another was the acceptance of nonconstructive procedures (definitions, proofs) as legitimate mathematical methods. Dedekind's great concern for teaching also influenced his mathematical thinking. His two very significant methodological innovations were the use of the axiomatic method and the institution of set-theoretic modes of thinking. See [14], [15].

To illustrate, compare Dedekind's and Kronecker's definitions of a field. (Recall that these two mathematicians developed essentially the same theory, at about the same time, but used entirely different methods and approaches.) Dedekind defined a field axiomatically (even if in a concrete setting), as the set of all real or complex numbers satisfying certain properties. Kronecker did not, in fact, define a field in the way *we* think of a definition. He *described* it, calling it a "domain of rationality": It consisted of all rational functions in the quantities $R', R'', R''', \ldots$ with integer coefficients $(Q(R', R'', R''', \dots))$ in current notation). He put no restrictions on these quantities—they could be indeterminates or roots of algebraic equations. In this sense, Kronecker's field concept is more general than Dedekind's. On the other hand, the elements $R', R'', R''', \dots$ had to be given explicitly, so that (for example) the field of all algebraic numbers would not be considered a domain of rationality, whereas it qualified as a field under Dedekind's definition. Here was a foreshadowing of the formalist-intuitionist controversy, to emerge in full force in the early decades of the 20th century.

The axiomatic method was just beginning to surface after 2000 years of near dormancy. Dedekind was instrumental in pointing to its mathematical power and pedagogical value. In this he inspired (among others) David Hilbert and Emmy Noether. His use of set-theoretic formulations (recall, for example, his definition of an ideal as the set of elements of a domain satisfying certain properties), including the use of the completed infinite—taboo at the time—preceded by about 10 years Cantor's seminal work on the subject.[14] Undeniably, it justifies the description of his 1871 memoir as "the 'birthplace' of the modern set-theoretic approach to the foundations of mathematics" [14, p. 9]. Edwards' tribute (especially as it comes from one who is a great admirer of Kronecker's approach to the subject) is fitting [14, p. 20]:

> Dedekind's legacy . . . consisted not only of important theorems, examples, and concepts, but of a whole *style* of mathematics that has been an inspiration to each succeeding generation.

Postscript

What of the three fundamental number-theoretic problems that gave rise to the development of algebraic number theory? In the 1871 memoir Dedekind gave a very satisfactory explanation, in terms of ideals, of Gauss' theory of binary quadratic forms, in particular of the composition of such forms. Specifically, with each quadratic form $f(x, y) = ax^2 + bxy + cy^2$ with discriminant $D = b^2 - 4ac$ Dedekind associated the ideal I of integers of the quadratic field $Q(\sqrt{D})$ generated by a and $(-b + \sqrt{D})/2$, that is, $I = \langle a, (-b + \sqrt{D})/2 \rangle$. He showed that this establishes a one-one correspondence between quadratic forms with discriminant D and ideals of the domain of integers of $Q(\sqrt{D})$, and that under this correspondence composition of forms corresponds to multiplication of ideals (see [1]). Although this result made Gauss' composition of forms transparent, it did not end interest in the (arithmetic) study of binary quadratic forms. In particular, the representation of integers by specific forms and the determination of the class number of forms are problems of current interest. See [10].

Various reciprocity laws beyond the biquadratic were formulated and proved in the 19th century, especially by Eisenstein, Kummer, and Hilbert. Kummer's work relied on his results on unique factorization of cyclotomic integers, Hilbert's initiated class-field theory (see [10], [17]). But these efforts were not entirely satisfactory. In fact, one of Hilbert's 23 problems presented at the 1900 International Congress of Mathematicians was to find a general reciprocity law. Artin solved the problem in the 1920s with his celebrated reciprocity law—a centerpiece of class-field theory. It establishes (in one of its incarnations, and roughly speaking) a connection between the Galois group of an abelian extension of an algebraic number field and the

---

[14]Several authors have commented on the methodological analogy between Dedekind's definitions of "ideal" and of "Dedekind cut". The definition of a cut, in fact, preceded that of an ideal. See [3], [13].

arithmetical properties of the ground field (see [10]). A far cry from the quadratic, cubic, and biquadratic reciprocity laws! See [21], [29].

Having instituted unique factorization in the domains D_p of cyclotomic integers, Kummer proved Fermat's Last Theorem for "regular" primes. (A prime p is *regular* if it does not divide the class number h of D_p; equivalently, if for any ideal I of D_p, if I^p is a principal ideal then so is I.) He then showed that all primes $p < 100$ (with the exception of 37, 59, and 67) are regular. Thus Fermat's Last Theorem was established for all primes $p < 100$ (about 10 years later Kummer disposed of the three remaining cases). Given that Gauss and Dirichlet failed to prove Fermat's theorem even for $n = 7$, this was quite an accomplishment. It showed the power of Kummer's theory of ideal numbers in cyclotomic domains. Much progress was made on Fermat's Last Theorem following Kummer's work (see [12] and [25]). In 1976, through the use of powerful computers along with powerful abstract mathematics, Fermat's Last Theorem was shown to hold for all primes $p < 125{,}000$ (see [25]). In another direction, as a corollary of his pioneering work on abelian varieties, Faltings showed in 1983 that *for each fixed $n > 2$ the equation $x^n + y^n = z^n$ has at most finitely many solutions* (See [4, pp. 41–42]). Wiles' very recent "proof" of Fermat's Last Theorem has its roots in Taniyama's 1955 conjecture about elliptic curves. It took 30 years to tie the conjecture to Fermat's Last Theorem. The high point of these developments was Ribet's 1987 proof that a special case of the Taniyama conjecture implies Fermat's Last Theorem. Wiles (apparently) devoted the past six years to proving this special case of Taniyama's conjecture. See [11], [26].

Emmy Noether used to say modestly of her work that it can already be found in Dedekind's. It would not be amiss to say that much of the subject matter of this paper originated in, or was inspired by, Gauss' *Disquisitiones Arithmeticae* of 1801. A new subject, algebraic number theory, came into being. It embodied fundamental concepts whose consequences reached far beyond number theory. Yet some of the major number-theoretic problems of that period are still with us today, two centuries later. Of course, the tools that have been brought to bear on their study are recent, and they are powerful. Gauss, Kummer, and Dedekind would likely have found them initially inscrutable, but in time wondrous. Mathematics is alive and well (if any evidence were needed).

REFERENCES

1. W. W. Adams and L. J. Goldstein, *Introduction to Number Theory*, Prentice-Hall, Englewood Cliffs, NJ, 1976.
2. M. F. Atiyah and I. G. Macdonald, *Introduction to Commutative Algebra*, Addison-Wesley Publishing Co., Reading, MA, 1969.
3. I. G. Bashmakova and A. N. Rudakov, The theory of algebraic numbers and the beginnings of commutative algebra, in *Mathematics in the 19th Century*, ed. by A. N. Kolmogorov and A. P. Yushkevich, Birkhäuser, Boston, 1992, pp. 86–135.
4. S. Bloch, The proof of the Mordell Conjecture, *Math. Intell.* 6:2 (1984), 41–47.
5. E. D. Bolker, *Elementary Number Theory: An Algebraic Approach*, W. A. Benjamin, New York, 1970.
6. N. Bourbaki, Historical note, in his *Commutative Algebra*, Addison-Wesley Publishing Co., Reading, MA, 1972, pp. 579–606.
7. H. Cohn, *Advanced Number Theory*, Dover Publications, Mineola, NY, 1980.
8. M. J. Collison, The origins of the cubic and biquadratic reciprocity laws, *Arch. Hist. Ex. Sc.* 17 (1977), 63–69.
9. G. Cornell, Review of two books on algebraic number theory, *Math. Intell.* 5:1 (1983), 53–56.
10. D. A. Cox, *Primes of the Form $x^2 + ny^2$: Fermat, Class Field Theory, and Complex Multiplication*, John Wiley & Sons, Inc., New York, 1989.
11. K. Devlin, F. Gouvêa, and A. Granville, Fermat's Last Theorem, a theorem at last, *Focus* 13:4 (1993), 3–4.

12. H. M. Edwards, *Fermat's Last Theorem: A Genetic Introduction to Algebraic Number Theory*, Springer-Verlag New York, 1977.

13. _____, The genesis of ideal theory, *Arch. Hist. Ex. Sc.* 23 (1980), 321–378.

14. _____, Dedekind's invention of ideals, *Bull. Lond. Math. Soc.* 15 (1983), 8–17.

15. _____, Mathematical ideas, ideals, and ideology, *Math. Intell.* 14:2 (1992), 6–19.

16. E. Grosswald, *Topics from the Theory of Numbers*, 2nd edition, Birkhäuser, Boston, 1984.

17. K. Ireland and M. Rosen, *A Classical Introduction to Modern Number Theory*, Springer-Verlag New York, 1982.

18. I. Kaplansky, Commutative rings, in *Proc. of Conf. on Commutative Algebra*, ed. by J. W. Brewer and E. A. Rutter, Springer-Verlag New York, 1973, pp. 153–166.

19. I. Kleiner, A sketch of the evolution of (noncommutative) ring theory, *L'Enseign. Math.* 33 (1987), 227–267.

20. M. Kline, *Mathematical Thought from Ancient to Modern Times*, Oxford University Press, New York, 1972.

21. E. Lehmer, Rational reciprocity laws, *Amer. Math. Monthly* 85 (1978), 467–472.

22. A. F. Monna, *L'algébrisation de la Mathématique: Réflexions Historiques*, Comm. Math. Inst. Rijksuniversiteit, Utrecht, 1977.

23. H. Pollard and H. G. Diamond, *The Theory of Algebraic Numbers*, 2nd edition, MAA, Washington, DC, 1975.

24. W. Purkert, Zur Genesis des abstrakten Körperbegriffs, I, II, NTM 8 (1971), 23–37 and 10 (1973), 8–20.

25. P. Ribenboim, *13 Lectures on Fermat's Last Theorem*, Springer-Verlag New York, 1979.

26. K. Ribet, Wiles proves Taniyama's conjecture; Fermat's Last Theorem follows, *Notices Amer. Math. Soc.* 40 (1993), 575–576.

27. H. Stark, *An Introduction to Number Theory*, MIT Press, Cambridge, MA, 1978.

28. A. Weil, Two lectures on number theory, past and present, *L'Enseign. Math.* 20 (1974), 87–110.

29. B. F. Wyman, What is a reciprocity law?, *Amer. Math. Monthly* 79 (1972), 571–586.

It had to be *e*!

(Sung to the tune of "It Had to be You!")

It had to be *e*,
Nonintegral *e*,
I looked around
Until I found
A base that would do.
To diff'rentiate
Or to integrate,
One that would not
Carry along
Some ugly weight.

Some bases I know
Are simpler to state,
A snap to invert,
Exponentiate,
But they wouldn't do.
For no other base can fit Math so well,
With all its digits I love it still!
It had to be *e*,
Irrational *e*,
It had to be *e*!

—WAYNE BARRETT
BRIGHAM YOUNG UNIVERSITY
PROVO, UT 84602

Do Estimates of an Integral Really Improve as *n* Increases?

SHERMAN K. STEIN
University of California
Davis, CA 95616-8633

One way to estimate $\int_0^1 f(x)\,dx$ is to partition the interval $[0,1]$ into n sections of equal lengths and use their right-hand endpoints to form the sum $\sum_{i=1}^n f(i/n)(1/n)$. Call this estimate E_n. If f is continuous, then E_n approaches the integral as n approaches infinity. A beginning calculus student or the author of a calculus text might ask, "As n increases, does the estimate get steadily closer to the integral?" In other words, is the approach monotonic?

A quick sketch can produce an example where E_1 equals the integral but E_2 does not. So the answer is no. However, if the derivative f' is nonnegative throughout the interval, then $E_1 \geq E_2 \geq \int_0^1 f(x)\,dx$ as another quick sketch will show. A similar conclusion holds if f' is nonpositive, in which case we have $E_1 \leq E_2 \leq \int_0^1 f(x)\,dx$. In short, if f' does not change sign in the interval $[0,1]$, then E_2 is at least as good an estimate as E_1, and similarly for any positive integers p and q, E_{pq} is at least as good as E_p.

Keeping the assumption that $f' \geq 0$, we next ask, "Is $E_2 \geq E_3$?" Once again, a diagram will show that the answer is no. And this automatically raises the next question: "For which general f with nonnegative first derivative, do we have $E_2 \geq E_3$?" At this point I talked to numerical analysts and looked in numerical analysis texts, but found no mention of the problem. Only after I had shown that $E_n \geq E_{n+1}$ for all n did I find that several mathematicians had studied this and related problems as early as 1929. I will first describe my own experience with the problem and its generalization and then the various techniques that others had employed in their solutions. Some of the arguments are elementary and could be sketched in a few minutes in a calculus class.

1. My Bout with the Problem

I began by using algebra, looking at the inequality $E_2 \geq E_3$, which, written out in full, is

$$\tfrac{1}{2}\left(f(1) + f(\tfrac{1}{2})\right) \geq \tfrac{1}{3}\left(f(1) + f(\tfrac{2}{3}) + f(\tfrac{1}{3})\right)$$

or, equivalently,

$$f(1) - 2f(\tfrac{2}{3}) + 3f(\tfrac{1}{2}) - 2f(\tfrac{1}{3}) \geq 0. \tag{1}$$

To avoid fractions, I replaced f by the function g defined by $g(x) = f(x/6)$, with domain $[0,6]$. Inequality (1) then read

$$g(6) - 2g(4) + 3g(3) - 2g(2) \geq 0. \tag{2}$$

What assumption about g in addition to the fact that g' is nonnegative, will imply that (2) holds?

To answer this, I used second differences. If g is a function and a and h are numbers, the expression $g(a+h)-g(a)$ is called a "first difference." Throughout h will be positive. The first difference of the first difference is called a "second difference," namely

$$[g(a+h+h)-g(a+h)]-[g(a+h)-g(a)]$$
$$=g(a+2h)-2g(a+h)+g(a).$$

The third difference, fourth difference, and so on are defined inductively, each being the first difference of the preceding one. The coefficients in the kth difference are the coefficients of the polynomial obtained by expanding $(x-1)^k$. In particular, the coefficient of the top term, $g(a+kh)$, is 1 and the coefficient of $g(a)$ is $(-1)^k$. An inductive argument shows that if g has a kth derivative then the kth difference equals $g^{(k)}(c)h^k$ for some number c in $[a, a+kh]$.

The expression "$g(6)-2g(4)$" in (2) suggested the second difference, $g(6) - 2g(4)+g(2)$. If g'' is nonnegative, so is that second difference. So rewriting (2) in the form

$$[g(6)-2g(4)+g(2)]+3[g(3)-g(2)]\geq 0$$

showed that if both g' and g'' are nonnegative, then $E_2 \geq E_3$. I recorded this as a theorem.

THEOREM 1.1. *If f' and f'' are nonnegative in $[0,1]$, then $E_2 \geq E_3$.*

In a similar manner, with $g(x)=f(x/12)$, I found that the inequality $E_3 \geq E_4$ reduced to

$$g(12)-3g(9)+4g(8)-3g(6)+4g(4)-3g(3)\geq 0.$$

The expression "$g(12)-3g(9)$" suggested the third difference,

$$g(12)-3g(9)+3g(6)-g(3),$$

and the inequality $E_3 \geq E_4$ became

$$[g(12)-3g(9)+3g(6)-g(3)]+4g(8)-6g(6)+4g(4)+2g(3)\geq 0,$$

and then

$$[g(12)-3g(9)+3g(6)-g(3)]+4[g(8)-2g(6)+g(4)]$$
$$+2[g(16)-g(3)]\geq 0.$$

So I had Theorem 1.2.

THEOREM 1.2. *If f', f'', and f''' are nonnegative in $[0,1]$, then $E_3 \geq E_4$.*

These results already suggested a conjecture and an approach to proving it. In any case, I looked at the next inequality, $E_4 \geq E_5$, which is equivalent to

$$g(20)-4g(16)+5g(15)-4g(12)+5g(10)-4g(8)+5g(5)-4g(4)\geq 0.$$
$$(3)$$

So I started with a fourth difference, and after some straightforward arithmetic found that (3) could be written in terms of differences:

$$[g(20)-4g(16)+6g(12)-4g(8)+g(4)]+5[g(15)-2g(12)+g(9)]$$
$$+5[g(10)-g(9)]+5[g(5)-g(4)]\geq 0.$$

So I had justified the next theorem.

THEOREM 1.3. *If f', f'', and $f^{(4)}$ are nonnegative in $[0, 1]$, then $E_4 \geq E_5$.*

The pattern was not what I expected: $f^{(3)}$ was missing. To get more information, I went on, using the same approach, obtaining the next two theorems.

THEOREM 1.4. *If $f', f'', f^{(3)}$, and $f^{(5)}$ are nonnegative in $[0, 1]$, then $E_5 \geq E_6$.*

THEOREM 1.5. *If $f', f'', f^{(3)}$ and $f^{(6)}$ are nonnegative in $[0, 1]$, then $E_6 \geq E_7$.*

These cases suggested that the hypotheses were very sensitive to n, but gave no hint of the general theorem. So I attacked the case $n = 7$, that is, $E_7 \geq E_8$. However, after several attempts, starting with a seventh difference, I could not express the inequality in terms of differences that met this critical condition: The lead coefficient of each difference is positive.

What to do? I felt that information about the higher derivatives must continue to imply the general inequality $E_n \geq E_{n+1}$. Moreover, it was clear that the problem was algebraic: how to represent vectors as linear combinations of certain prescribed vectors, namely, the differences, with nonnegative coefficients.

As I looked over the settled cases, I saw that f' and f'' appeared in all of them. Could it be that $f^{(3)}$, $f^{(4)}$, $f^{(5)}$, and $f^{(6)}$ were just superfluous distractions. I had been using the greedy algorithm, where a high-order difference mimicked the first two terms of the typical inequality. Maybe I should switch to a timid approach. However, I knew that I couldn't get by with just first differences, for, if I could, then that would show that if f' were nonnegative, then $E_2 \geq E_3$. So I would have to exploit at least second differences. The first test case was the inequality $E_3 \geq E_4$, which amounted to

$$g(12) - 3g(9) + 4g(8) - 3g(6) + 4g(4) - 3g(3) \geq 0. \tag{4}$$

Mimicking only $g(12)$ and a bit of the next summand, I started by writing (4) as

$$[g(12) - 2g(9) + g(6)] - g(9) + 4g(8) - 4g(6) + 4g(4) - 3g(3) \geq 0. \tag{5}$$

But $g(9)$ has the negative coefficient -1. The next second difference to use in rewriting (5) would be $-[g(9) - 2g(8) + g(7)]$, which would be nonpositive if f'' were nonnegative. That would be useless. So I had to settle the question, "Can the left side of (4) be expressed as the sum of first and second differences with positive coefficients?" The emphasis is on the world "positive".

Maybe my timid algorithm wasn't timid enough. Perhaps I had tried to do too much by mimicking $g(12)$ and part of the next term, $-3g(9)$. I wondered whether I could avoid negative coefficients by being less demanding.

I could start instead with $g(12) - 2g(10.5) + g(9)$, which would then leave me with $2g(10.5)$ as the first remaining term to mimic. However, the noninteger argument 10.5 would open a can of worms and went against all my instincts. Surely, if (4) could be represented the way I wanted, it should not require noninteger arguments.

So, instead, I took a shot in the dark, beginning with the second difference $g(12) - 2g(11) + g(10)$. It was very timid, since I didn't try to go after $g(9)$. However, it was daring, since I was introducing $g(11)$ and $g(10)$, which weren't present in (4) and therefore made me a bit uneasy. However, it did give a positive coefficient as the next term to be mimicked since (4) had become

$$[g(12) - 2g(11) + g(10)] + 2g(11) - g(10) - 3g(9) + 4g(8)$$
$$- 3g(6) + 4g(4) - 3g(3) \geq 0. \tag{6}$$

Mimicking $2g(11)$, I used the timid $2g(11) - 4g(10) + 2g(9)$. So (4) now had become

$$[g(12) - 2g(11) + g(10)] + 2[g(11) - 2g(10) + g(9)] + 3g(10) - 5g(9)$$
$$+ 4g(8) - 3g(6) + 4g(4) - 3g(3) \geq 0.$$

The term $3g(10)$, again happily with a positive coefficient, suggested using $3[g(10) - 2g(9) + g(8)]$, and I operated on (6) to get

$$[g(12) - 2g(11) + g(10)] + 2[g(11) - 2g(10) + g(9)]$$
$$+ 3[g(10) - 2g(9) + g(8)]$$
$$+ g(9) + g(8) - 3g(6) + 4g(4) - 3g(3) \geq 0.$$

Since $g(9)$ has a positive coefficient, I could use $g(9) - 2g(8) + g(7)$. Two additional steps then yielded

$$[g(12) - 2g(11) + g(10)] + 2[g(11) - 2g(10) + g(9)]$$
$$+ 3[g(10) - 2g(9) + g(8)]$$
$$+ [g(9) - 2g(8) + g(7)]$$
$$+ 3[g(8) - 2g(7) + g(6)] + 5[g(7) - 2g(6) + g(5)]$$
$$+ 4[g(6) - 2g(5) + g(4)] + 3[g(5) - g(3)] \geq 0. \qquad (7)$$

So I had expressed (4) in terms of first and second differences, all with positive coefficients. As a consequence, I saw that Theorem 1.2 was true without any mention of $f^{(3)}$.

This timid approach was as timid as could be. At each step the argument went down only by 1, with a second difference starting at $g(12)$, then at $g(11)$, then at $g(10)$, and so on. This approach was an automatic procedure, an algorithm. All would be settled if the algorithm gave only positive coefficients for the second differences (and for the first differences that appear at the end).

I tested the algorithm for $n = 4, 5, 6$, and the pesky case, 7. It went through smoothly, and suggested that the theorem to prove had a fixed hypothesis: If f' and f'' are nonnegative in $[0, 1]$, then $E_n \geq E_{n+1}$.

I didn't bother to check the case $n = 8$. The pattern was so uniform that all that remained was to describe it. The proof would then follow easily.

As I looked back at the calculations one thing struck me. The function g played no role in the manipulation. In that case it should be possible to replace differences by the polynomials they resemble. This suggested that $g(a + 2h) - 2g(a + h) + g(a)$ could be represented by $x^{a+2h} - 2x^{a+h} + x^a = x^a(x^h - 1)^2$. In that case, $c[g(a + 2h) - 2g(a + h) + g(a)]$ would correspond to $cx^a(x^h - 1)^2$. For instance, the identity

$$g(6) - 2g(4) + 3g(3) - 2g(2) = [g(6) - 2g(4) + g(2)] + 3[g(3) - g(2)]$$

reads, in the language of polynomials,

$$x^6 - 2x^4 + 3x^3 - 2x^2 = [x^6 - 2x^4 + x^2] + 3[x^3 - x^2] = x^2(x^2 - 1)^2 + 3x^2(x - 1).$$

Since polynomials form a structure with addition and multiplication, they might provide a convenient arena in which to carry out the bookkeeping. This approach did work out and gave a much more direct proof of the following theorem.

THEOREM 1.6. *Let the function f be defined on* $[0, 1]$ *and be twice differentiable. If* f' *and* f'' *are nonnegative, then* $E_n \geq E_{n+1}$.

In outline, the proof goes like this. First note that a second difference with $h > 1$ can be expressed as the sum of second differences with $h = 1$, all with the same sign. For instance,

$$g(6) - 2g(4) + g(2) = [g(6) - 2g(5) + g(4)] + 2[g(5) - 2g(4) + g(3)]$$
$$+ [g(4) - 2g(3) + g(2)].$$

To see that this holds in general, note that

$$cx^2(x^h - 1)^2 = cx^a(x^{h-1} + \cdots + 1)^2(x - 1)^2.$$

Similarly, we may restrict h to 1 in our first differences.

The expression $E_n - E_{n+1}$ then is represented by a polynomial $A(x)$ of degree $n(n + 1)$. Now

$$A(x) = Q(x)(x - 1)^2 + c(x - 1) + d, \tag{8}$$

for a unique polynomial $Q(x)$ in $\mathbb{Z}[x]$, and unique elements $c, d \in \mathbb{Z}$. At this point the proof is straightforward: Compute c, d, and $Q(x)$ and show that c, d, and the coefficients of $Q(x)$ are nonnegative. The computations are not hard since division by $x - 1$ is simple: If $A(x) = B(x)(x - 1) + d$, the coefficients of $B(x)$ are convenient sums of coefficients of $A(x)$, and $d = A(1)$. Then division of $B(x)$ by $x - 1$ completes the argument; the only nonroutine step is proving that the coefficients of the quotient are nonnegative.

Theorem 1.6 can be strengthened. Applying the theorem to the function $-f$ shows that if f' and f'' are nonpositive, then E_{n+1} is again at least as good an estimate of $\int_0^1 f(x)\, dx$ as E_n is. A proof similar to that of Theorem 1.6 shows that if f' is nonnegative and f'' is nonpositive the same conclusion holds. This is all summarized in the next theorem. In this case, subtract the second differences with negative coefficients, starting at the *smallest* argument rather than at the largest one.

THEOREM 1.7. *Let f be defined on* $[0, 1]$ *and be twice differentiable. If* f' *and* f'' *do not change sign, then* $E_n \leq E_{n+1} \leq \int_0^1 f(x)\, dx$ *or* $E_n \geq E_{n+1} \geq \int_0^1 f(x)\, dx$ *for all n.*

After obtaining Theorem 1.7, I played a little with trapezoidal estimates and conjectured that if neither $f^{(2)}$ nor $f^{(3)}$ changes sign, then these estimates approach the integral monotonically. Moreover, an argument similar to the one for the Riemann sum estimates, but using third and second differences, looked like it would work but be quite messy. It also looked as if neither $f^{(4)}$ nor $f^{(5)}$ changes sign, then the Simpson estimates also approach the integral monotonically. A pattern emerged: If an approximation method was exact for polynomials of degree at most k then it would be monotonic if $f^{(k+1)}$ and $f^{(k+2)}$ don't change sign. Before attempting to prove such a broad conjecture, I thought it prudent to be sure it was not already a theorem, perhaps a century old but forgotten. In response to inquiries, T. Rivlin mentioned a paper of D. J. Newman [4] and one of his own that examine the monotonicity of the midpoint method [6], both published in 1974. Following this lead, I found that "monotonicity" questions had been investigated as far back as 1929. I will describe two elementary approaches in detail and state Newman's result, which confirmed my conjecture for a broad class of estimation techniques.

2. Reduction to Simpler Functions

The earliest appearance of the monotonicity problem in print was in 1950 in a paper of P. Turán [11]. In connection with the study of the zeros of the Legendre polynomials, he cited "some unpublished theorems of Fejér, which I mention here with his kind permission; he found his results during the winter term of the academic year 1928/1929." Theorem 2.1 below is one of the four results of Fejér, the others obtained by other combinations of "nonincreasing" or "nondecreasing" with "concave" or "concave down." In 1961 Turán and Szego published a proof in a joint paper [10]. Their approach depends on an observation that substantially reduces the class of functions to be examined.

Consider the class C of decreasing functions $f(x)$ on $[0, 1]$ that are convex ("concave up") in the sense that they lie below their chords. For convenience assume $f(1) = 0$.

This class includes the special functions

$$f_r(x) = \begin{cases} r - x & \text{for } 0 \leq x \leq r \\ 0 & \text{for } r \leq x \leq 1, \end{cases} \tag{9}$$

for all $0 \leq r \leq 1$.

Any function in C can be uniformly approximated (from below) by a suitable linear combination of the functions $f_r(x)$ with *positive* coefficients. FIGURE 1 illustrates why this is the case.

Starting at $(1, 0)$ move to the left to $(r_1, 0)$ until the curve is ε above the x-axis. From $(r_1, 0)$ draw the tangent to the curve and continue it to the left until it is ε below the curve, at a point with x coordinate r_2. Continue constructing such segments of tangents until meeting the y-axis, obtaining a finite sequence of numbers $r_1, r_2, \ldots, r_k$. Then there are positive constants $c_1, \ldots, c_k$ such that $|f(x) - \sum_{i=1}^{k} c_i f_{r_i}(x)| \leq \varepsilon$ for all x in $[0, 1]$. This is the key to Theorem 2.1, which had been stated by Fejér.

Of the four classes of functions to consider, we take only the case $f(x)$ "decreasing" and "concave up."

The theorem concerns two point systems in $[0, 1]$, namely

$$0 = x_0 < x_1 < \cdots < x_n < x_{n+1} = 1$$

and

$$0 = y_0 < y_1 < \cdots < y_{n+1} < y_{n+2} = 1.$$

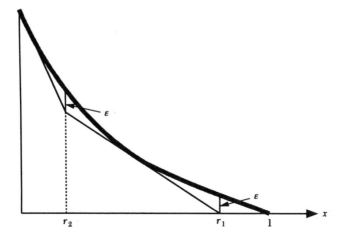

FIGURE 1

The first divides $[0,1]$ into $n+1$ sections and the second into $n+2$ sections. Moreover, we assume that the point systems interlace each other, that is,

$$x_i < y_{i+1} < x_{i+1},$$

$i = 0, 1, \ldots, n$, as shown in FIGURE 2.

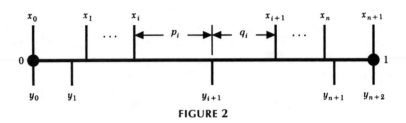

FIGURE 2

Let $p_i = y_{i+1} - x_i$ and $q_i = x_{i+1} - y_{i+1}$, $i = 0, 1, \ldots, n$, as shown in FIGURE 2. (If both the x_i's and the y_i's are equally spaced, they form an interlaced system. A simple computation shows that in this case $p_0 \geq p_1 \geq p_2 \geq \cdots \geq p_n$.) Let

$$S_n(f) = \sum_{i=0}^{n} f(x_i)(x_{i+1} - x_i) \quad \text{and} \quad S_{n+1}(f) = \sum_{i=0}^{n+1} f(y_i)(y_{i+1} - y_i),$$

the "left-hand" Riemann sums formed with these two subdivisions. We take this case since it is the one treated by Szegó and Turán. They prove the following theorem, which they credit to Fejér.

THEOREM 2.1. *In order that $S_{n+1}(f) \geq S_n(f)$ holds for a fixed n and all nonincreasing concave up functions, it is necessary and sufficient that*

$$\sum_{i=0}^{j} q_i(p_{i+1} - p_i) \leq 0 \tag{10}$$

for each j, $0 \leq j \leq n$.

Sketch of proof (Assume that $f(1) = 0$). Since any nonincreasing concave up function in $[0,1]$ can be uniformly approximated by the special functions (9), we restrict our attention to these functions. The idea is to show that the set of $n+1$ inequalities (10) is equivalent to the inequalities

$$S_{n+1}(f_r) \leq S_n(f_r) \tag{11}$$

for all r, $0 \leq r \leq 1$. To do this, first evaluate both sides of (11).

Given r, $0 \leq r \leq 1$, define j by

$$x_j \leq r < x_{j+1}.$$

If $r = 1$ define j to be $n + 1$. Then define a by the equation

$$r = x_j + a.$$

Thus, if $j > 1$, we have

$$r = p_0 + q_0 + p_1 + q_1 + \cdots + p_{j-1} + q_{j-1} + a,$$

and, if $j = 0$, then $r = a$. Also we have

$$0 \leq a < p_j + q_j.$$

We first evaluate $S_n(f_r)$ with the aid of FIGURE 3.

If $j = 0$, $S_n(f_r) = ax_1$. If $1 \leq j \leq n$, then $S_n(f_r)$ can be viewed as the sum of the area of the large triangle of area $r^2/2$, some smaller triangles above it, and a trapezoid. We obtain

$$S_n(f_r) = \frac{r^2}{2} + \frac{1}{2} \sum_{i=0}^{j-1} (p_i + q_i)^2 + \left(p_j + q_j - \frac{a}{2}\right)a,$$

which gives

$$S_n(f_r) = \frac{r^2 - a^2}{2} + \frac{1}{2} \sum_{i=0}^{j-1} p_i^2 + \sum_{i=0}^{j-1} p_i q_i + \frac{1}{2} \sum_{i=0}^{j-1} q_i^2 + (p_j + q_j)a. \qquad (12)$$

The evaluation of $S_{n+1}(f_r)$ is similar but splits into two cases, $a \leq p_j$ and $a > p_j$. We leave it for the reader to treat these two cases and complete the proof.

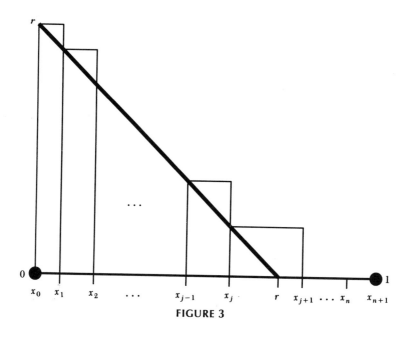

FIGURE 3

3. Jensen's Inequality

In response to my question, A. Pinkus sent me a different approach in May 1992, which exploits the fact that by definition, a convex function lies below its chords. Analytically, this condition reads as follows.

Let $a \leq c \leq b$. There are unique numbers p and q such that $0 \leq p, q \leq 1$, $p + q = 1$, and $c = pa + qb$. Then a function is convex on the interval $[a, b]$ if, and only if,

$$f(pa + qb) \leq pf(a) + qf(b) \qquad (13)$$

for all such p and q. (The special case when c is the midpoint of $[a, b]$, hence $p = 1/2 = q$, was used in Section 1.)

THEOREM 3.1. *Let the function f defined on $[0, 1]$ be convex and nonincreasing. Then, in the notation of Theorem 1.6, $E_n \leq E_{n+1}$.*

Sketch of proof. Note that $i/(n+1) < i/n < (i+1)/(n+1)$, and we have

$$\frac{i}{n} = \frac{n-i}{n} \cdot \frac{i}{n+1} + \frac{i}{n} \cdot \frac{i+1}{n+1}.$$

By (13)

$$f\left(\frac{i}{n}\right) \le \frac{n-i}{n} f\left(\frac{i}{n+1}\right) + \frac{i}{n} f\left(\frac{i+1}{n+1}\right). \tag{14}$$

Applying (14) to each summand in $\sum_{i=1}^{n}(1/n)f(i/n)$ will complete the proof.

4. The Generalization to All Equi-spaced Newton-Cotes Estimates

The right-hand, left-hand, midpoint, trapezoidal, and Simpson's methods are all special cases of *Newton-Cotes quadratures*. These quadratures estimate $\int_0^1 f(x)\,dx$ as follows.

Let k be a nonnegative integer and $x_0 < x_1 < \cdots < x_k$ be $k+1$ points in $[0,1]$. (In applications, usually $x_0 = 0$, $x_k = 1$ and the points are equally spaced.) Let $p(x)$ be the unique polynomial of degree at most k such that $p(x_i) = f(x_i)$, $0 \le i \le k$. Then the Newton-Cotes estimate of $\int_0^1 f(x)\,dx$ is $\int_0^1 p(x)\,dx$. Note that the estimate is exact when $f(x)$ is a polynomial of degree $\le k$. Equivalently, one chooses fixed "weights" $a_0, a_1, \ldots, a_k$ such that the estimate $\sum_{i=0}^{k} a_i f(x_i)$ equals $\int_0^1 f(x)\,dx$ for $f(x) = x^0, x^1, \ldots, x^k$.

For $k = 0$ we get the left-hand, right-hand, or midpoint estimates according as x_0 is chosen to be 0, $1/2$, or 1, respectively. For $k = 1$, $x_0 = 0$, $x_1 = 1$ we get the trapezoidal estimate (with a single trapezoid) and for $k = 2$, $x_0 = 0$, $x_1 = 1/2$, $x_2 = 1$, Simpson's estimate (with a single parabola).

Consider a fixed Newton-Cotes method. Let n be a positive integer and break the interval $[a,b]$ into n sections of equal lengths. Estimate the integral of $f(x)$ over each of the n sections by a single application of that method. The sum of these n "local" estimates is called the *composite estimate* of $\int_a^b f(x)\,dx$, and is denoted $L(n)$. (The trapezoidal and Simpson's methods presented in freshman calculus are examples of composite estimates.)

In 1974 D. J. Newman [4] obtained the following general theorem.

THEOREM 4.1. *Let k be a positive integer and consider the Newton-Cotes procedure based on k sections of equal length. Let L_n denote the corresponding composite estimate of $\int_a^b f(x)\,dx$. Assume that $f^{(k+1)}$ and $f^{(k+2)}$ are continuous and do not change sign in $[a,b]$. Then the estimates $L_1, L_2, \ldots, L_n, \ldots$ approach $\int_a^b f(x)\,dx$ monotonically.*

For the proof, which depends on advanced calculus, we refer the reader to the paper itself.

5. Questions That Remain

In Newman's theorem the underlying Newton-Cotes method is based on $k+1$ equally spaced points $x_0, x_1, \ldots, x_k$, with x_0 and x_k being the ends of the interval. It does not include the midpoint method, for example. Is Theorem 4.1 a special case of a more general theorem? Is it true that if a Newton-Cotes procedure is exact for polynomials of degree k but not of degree $k + 1$ and neither $f^{(k+1)}$ nor $f^{(k+2)}$ changes sign, then the estimates L_n converge monotonically?

Might the following known generalization of Jensen's inequality, stated as Theorem 5.2, do the trick, along the lines of the proof of Theorem 3.1?

THEOREM 5.2. *Let $f(x)$ have a nonnegative nth derivative everywhere and let $a_0, a_1, \ldots, a_n$ be $n + 1$ distinct real numbers. Then*

$$\sum_{i=0}^{n} \frac{f(a_i)}{\prod_{\substack{0 \le j \le n \\ j \ne i}} (a_i - a_j)} \ge 0.$$

The case $n = 2$ is Jensen's inequality.

Could the technique of reducing the general case to a special case of functions, as in the proof of Theorem 2.1, be generalized? For instance, is there a way of approximating functions for which $f^{(2)}$ and $f^{(3)}$ don't change sign, by a sum of second-degree polynomials?

What about the algebraic approach? It provides an algorithm for testing any specific case as long as the points of subdivision are rational. But lurking behind this algebraic equation are several concerning the group $\mathbb{Z}^n$. For instance, *when is an element in $\mathbb{Z}^n$ expressible as a linear combination of "kth differences" with nonnegative coefficients?* (An element in $\mathbb{Z}^n$ of the form $(0, \ldots, 0, 1, k, \ldots, k, 1, 0, \ldots, 0)$, where the nonzero entries are the numbers $(-1)^i \binom{k}{i}$, corresponds to a kth difference with difference 1.) Of interest in our problem is the question, when is an element in $\mathbb{Z}^n$ expressible as a linear combination of $(k + 1)$st differences with nonnegative (nonpositive) coefficients and $(k + 2)$nd differences with nonnegative (nonpositive) coefficients?

After all this, a calculus student might ask, "But how does a *single* application of the Newton-Cotes method for k sections compare with a *single* application for $k + 1$ sections? For instance, how does the trapezoidal estimate compare with Simpson's method?"

This question was considered by H. Brass [2] in 1978. Let Q_m be the Newton-Cotes estimate of $\int_a^b f(x)\, dx$ based on the $m + 1$ equally spaced points $x_0 = a$, $x_1, x_2, \ldots,$ $x_m = b$. He proved that if $f^{(2k)}(x) \ge 0$, then $Q_{2k-1} \ge Q_{2k} \ge Q_{2k+1}$. Thus, if all even order derivatives are nonnegative then the various Newton-Cotes estimates approach $\int_a^b f(x)\, dx$ monotonically.

Of course, the student then asks, "But how do the composite estimates based on trapezoids compare with a single Simpson estimate?" If T_k denotes the estimate based on k trapezoids of equal width and S_k denotes Simpson's estimate based on k parabolas of equal width, how does T_k compare with S_1? Using the algebraic approach, one may show that if $f^{(2)}(x) \ge 0$, then $T_i \ge S_1$, $i = 1, 2,$ and 3.

This same persistent student might ask, "In the trapezoidal method the weights are $1/2$ and $1/2$. In Simpson's method they are $1/6, 4/6, 1/6$. Are they always positive in any Newton-Cotes procedure with equally spaced points?" Intuition may suggest "yes", and it would be right through the first seven cases. However, some of the weights are negative in the case of eight sections (that is, nine equally spaced points). With nine sections, all weights are positive. If more sections are used, then approximately half the weights are negative. (See [12].) The cancellation effect of positive and negative terms increases the importance of roundoff errors and is one of the reasons that higher order Newton-Cotes procedures are impractical.

Not only do some of the weights become negative in high order Newton-Cotes estimates, but some become very large. Though a_0 and a_k are asymptotic to $1/(k \log k)$ and a_1 and a_{k-1} are asymptotic to $1/(\log k)^2$, all the rest of the weights become arbitrarily large in absolute values as $k \to \infty$. For instance, for even k, $k = 2r$, a_r is asymptotic to

$$\frac{(-1)^{r-1} 4^r}{\sqrt{\pi} \, (\log 2r)^2}.$$

(See [13].)

Other questions may occur to the reader. (For the monotonicity of the Gauss or Romberg estimates see [2] and [9]. For the relation to convexity cones see [1] and [14].)

In any case, when we assure our calculus students that "the estimates get better as we subdivide finer and finer," maybe we should add the caveat, "as a general trend."

Acknowledgment. I wish to thank Anthony Barcellos for preparing the illustrations (using Cohort software).

REFERENCES

1. D. Amir and Z. Ziegler, Convexity cones and their duals, *Pacific J. Math.* 27 (1968), 425–440.
2. H. Brass, Monotonie bei den Quadraturverfahren von Gauss and Newton-Cotes, *Numer. Math.* 30 (1978), 349–354.
3. D. Kincaid and W. Cheney, *Numerical Analysis*, Brooks/Cole Publishing Co., Belmont, CA, 1990, p. 478.
4. D. J. Newman, Monotonicity of quadrature approximations, *Proc. Amer. Math. Soc..* 42 (1974), 251–257.
5. A. Pinkus, letter of May 11, 1992.
6. T. J. Rivlin, The Lebesgue constants for polynomial interpolation, in *Functional Analysis and Its Applications*, Ed. H. G. Garnir et al., Lecture Notes in Mathematics, 399 (1974), 422–437 (in particular, 428–429).
7. F. Stenger, Bounds on the error of Gauss-type quadratures, *Num. Math.* 8 (1966), 150–160.
8. J. Stoer and R. Bulirsch, *Introduction to Numerical Analysis*, Springer-Verlag New York, Inc., New York (1979), 123–125.
9. T. Ström, Monotonicity in Romberg quadrature, *Math. Comp.* 26 (1972), 461–465.
10. G. Szegö and P. Turan, On the monotone convergence of certain Riemann sums, *Publ. Math. Debrecen* 8 (1961), 326–335.
11. P. Turán, On the zeros of the polynomials of Legendre, *Casopis Pest Mat. Fys.* 75 (1950), 113–122, (MR Vol 12), p. 824).
12. J. V. Uspensky, On the expansions of the remainder in the Newton-Cotes formula, *Trans. Amer. Math. Soc.* 17 (1935), 381–396.
13. _____, Sur les valeurs asymptotiques des coefficients de Cotes, *Bull. Amer. Math. Soc.* 31 (1925), 145–156.
14. Z. Ziegler, Generalized convexity cones, *Pacific J. Math.* 27 (1966), 561–580.

The Box Problem:
To Switch or Not to Switch

STEVEN J. BRAMS
Department of Politics
New York University
New York, NY 10003

D. MARC KILGOUR
Wilfrid Laurier University
Waterloo, Ontario, Canada N2L 3C5

1. Introduction

Imagine that you are shown two identical boxes. You know that one of them contains b and the other $2b$. Picking one at random and opening it, you must decide whether to keep it (and its contents), or exchange it for the other box.

Suppose the box you pick contains $100. You think to yourself:

Either I picked the b box (with probability $1/2$), so the other box contains $200, or I picked the $2b$ box (with probability $1/2$), so the other box contains $50. Therefore, my expected value if I decide to switch is $(1/2)\$200 + (1/2)\$50 = \$125$. Given that my objective is to maximize expected value, I should make the switch.

But then you think:

Can this be right? If I had found that the box I picked contained x, the same calculation would have told me that the expected value of switching would be $1.25x$, so I would have decided to switch no matter what I discovered in the first box.

It seems paradoxical that it would always be better to switch to the second box, no matter how much money you found in the first one.

In fact, an obvious weakness of the expected-value argument is that it does not make use of the information you gained when you opened the first box. What is needed to determine whether or not switching is worthwhile is some prior notion of the amount of money to be found in each box. In particular, knowing the prior probability distribution would enable you to calculate whether the amount found in the first box is less than the expected value of what is in the other box—and, therefore, whether a switch is profitable.

We shall show that a switch is justified if, and only if, the conditional probability that the first box contains the larger amount, once its contents are observed (this is the condition), is less than $2/3$. Surprisingly, there are both discrete and continuous distributions for which this inequality never fails: A switch is *always* called for, no matter how much you find in the first box. Furthermore, it is always possible to find a sufficiently small amount in the first box that makes switching your rational choice.

Whether switching is occasionally or always rational depends on the prior distribution, which might depend on either objective or subjective factors, or some combination. There might, for example, be an objective record of how players have fared in the past from switching, comparable to the records of investors who switched in and out of different kinds of investments at different times. Or a person might have a subjective feeling, based on no hard evidence, that directs his or her choice.

We conclude by discussing the relationship of the box problem to utility theory and the St. Petersburg paradox. Implications of our findings for determining when envy is rational are noted.

2. The Exchange Condition

Assume each box contains some positive amount. Let L be a random variable representing the larger amount, and let S represent the smaller amount. Because $S = L/2$, a prior distribution for L defines a prior distribution for S such that

$$\Pr(L \leq x) = \Pr(S \leq x/2), \ 0 < x < \infty.$$

You begin by picking one of the two boxes at random. Call this box B1, and denote its contents by the random variable X. Let the other box be B2, and denote its contents by Y. Note that

$$\Pr(X = L) = \Pr(X = S) = 1/2.$$

Now you observe the contents of B1, $X = x$. A switch will be profitable if, and only if, the expected value of the amount in the other box, B2, is greater than x:

$$E(Y|X = x) = (x/2)\Pr(X = L|X = x) + (2x)\Pr(X = S|X = x) > x, \text{ or}$$
$$(1/2)\Pr(X = L|X = x) + (2)\Pr(X = S|X = x) > 1. \tag{1}$$

But if $X \neq L$, then $X = S$; therefore,

$$\Pr(X = S|X = x) = 1 - \Pr(X = L|X = x). \tag{2}$$

Substituting (2) into (1),

$$(1/2)\Pr(X = L|X = x) + 2[1 - \Pr(X = L|X = x)] > 1, \text{ or}$$
$$(3/2)\Pr(X = L|X = x) < 1,$$

so an exchange is profitable if, and only if,

$$\Pr(X = L|X = x) < 2/3. \tag{3}$$

Therefore, you should switch from B1 to B2 if, and only if, the conditional probability that you picked the L box is less than $2/3$, based on the prior probability distribution and given the observed $X = x$. We call (3) the *General Exchange Condition*—you should switch if, and only if, it is true.

3. Discrete Distributions

We first consider the case in which X, L, and S are discrete random variables. Assume there exists a fixed $m > 0$ such that the amount in any box must equal $\$2^k m$ for some integer k (which may, of course, be nonpositive). The 2^k factor reflects the fact that the next-smaller and next-larger amounts are, respectively, half and double the amounts in the box chosen.

Define

$$\Pr(X = 2^k m|X = L) = p_k$$

for $k = \ldots -1, 0, 1 \ldots$. We require that $\{\ldots, p_{-1}, p_0, p_1, \ldots\}$ defines a probability distribution, so assume $p_k \geq 0$ and $\sum_{k=-\infty}^{\infty} p_k = 1$. Notice that

$$\Pr(X = 2^k m|X = S) = \Pr(X = 2^{k+1} m|X = L) = p_{k+1}.$$

Now suppose you observe $X = 2^k m$ for some k. Because X has been observed,

$$\Pr(X = 2^k m | X = L) + P(X = 2^k m | X = S) = p_k + p_{k+1} > 0. \tag{4}$$

It follows from Bayes' Theorem that

$$\Pr(X = L | X = 2^k m)$$

$$= \frac{\Pr(X = 2^k m | X = L)\Pr(X = L)}{\Pr(L = 2^k m | X = L)\Pr(X = L) + \Pr(S = 2^k m | X = S)\Pr(X = S)}$$

$$= \frac{p_k}{p_k + p_{k+1}}. \tag{5}$$

The General Exchange Condition (3) now becomes

$$\frac{p_k}{p_k + p_{k+1}} < \frac{2}{3}, \quad \text{or} \quad p_k < 2p_{k+1}, \tag{6}$$

which we shall refer to as the *Exchange Condition for Discrete Distributions*. It says that whatever you observe, you should exchange if, and only if, the (unconditional) probability that you picked the L box (p_k) is less than twice the (unconditional) probability that you picked the S box (p_{k+1}). In other words, if the probability you picked the L box (as opposed to the S box) is large enough, keep this box.

We next give four discrete distributions where Exchange Condition (6) is satisfied at least some of the time:

1. Exchange Condition (6) usually satisfied. $m = 1$; $p_k = 1/4$ for $k = 0, 1, 2, 3$; $p_k = 0$ for all other k. Thus, the L box contains $1, $2, 4, or $8 equiprobably. The Exchange Condition is satisfied for $x = 1/2$, 1, 2, and 4 but not for $x = 8$. This condition holds with probability $7/8$, because for (6) to fail, both $X = L$ and $L = 8$ must occur, which has probability $(1/2)(1/4) = 1/8$.

2. Exchange Condition (6) usually not satisfied. $m = 1$; $p_k = 2/3^k$ for $k = 1, 2, 3, \ldots$; $p_k = 0$ for all other k. Thus, the L box contains $2 with probability $2/3$, $4 with probability $2/9$, $8 with probability $2/27$, etc. The Exchange Condition holds only if $x = 1$, which occurs when box $X = S$ and $S = 1$, and has probability $(1/2)(2/3) = 1/3$.

3. Exchange Condition (6) always satisfied. $m = 1$; $p_k = 2^{k-1}/3^k$ for all $k = 1, 2, 3, \ldots$; $p_k = 0$ for other k (see [4] for a similar example). Thus, the L box contains $2 with probability $1/3$, $4 with probability $2/9$, $8 with probability $4/27$, etc. The Exchange Condition is always satisfied.

Note that example 3 has a "floor value," $x = 1$, for which the decision to exchange is certain to be better, because it is the lowest possible amount in a box. The next example has no such floor, but still the Exchange Condition is always satisfied.

4. Exchange Condition (6) always satisfied, and no floor. $m = 1$; $p_{-k} = p_k = (1/7)(2/3)^{k-1}$ for $k = 1, 2, 3, \ldots$; $p_0 = 1/7$. Here the L box contains $(1/2)$, $1, and $2 with probability $1/7$ each, $(1/4)$ and $4 with probability $2/21$ each, etc. Again, the Exchange Condition is always satisfied.

Note that example 4 is a distribution that forms a plateau at $x = 1/2$, 1, and 2, falling off in each direction as $x \to 0$ and $x \to \infty$. The rate of decrease as x increases is slow enough, however, that (6) is always satisfied.

We now show

THEOREM 1. *For any probability distribution* $\{\ldots, p_{-1}, p_0, p_1, \ldots\}$, *Exchange Condition* (6) *must hold for at least one value of* k.

Proof. Suppose, to the contrary, that $p_{k-1} \geq 2p_k$ for $k = \ldots -1, 0, 1, \ldots$. Choose a k such that $p_k > 0$. If $n < k$, then $p_n \geq 2^{k-n} p_k$. But this means that $p_n \to \infty$ as $n \to -\infty$, contradicting the assumption that p_n is a probability.

Thus, no matter what the discrete distribution is, Exchange Condition (6) holds for some values of x. Indeed, it may hold for all values of x, as examples 3 and 4 illustrate; switching is always profitable in such cases.

4. Continuous Distributions

We next consider the case in which X, L, and S are absolutely continuous random variables. Because

$$\Pr(S \leq x) = \Pr(L \leq 2x),$$

the cumulative distribution functions of S and L are related by

$$F_S(x) = F_L(2x). \tag{7}$$

Differentiating (7) demonstrates that the probability density functions are related by

$$f_S(x) = 2f_L(2x).$$

Now suppose that you pick B1 and find that X satisfies $x \leq X \leq x + dx$. The conditional probability that B1 is the L box, given this observation, is analogous to the probability given by (5) in the discrete case:

$$\Pr(X = L | x \leq X \leq x + dx) = \frac{f_L(x)\,dx}{f_L(x)\,dx + f_S(x)\,dx} = \frac{f_L(x)}{f_L(x) + 2f_L(2x)}.$$

The continuous analogue of (3) is, therefore,

$$\frac{f_L(x)}{f_L(x) + 2f_L(2x)} < \frac{2}{3},$$

which is equivalent to the following *Exchange Condition for Continuous Distributions*:

$$f_L(x) < 4f_L(2x). \tag{8}$$

Thus, an exchange is profitable if, and only if, the (unconditional) density of L at x, the observed value of X, is less than four times the (unconditional) density of L at $2x$. In other words, the value of $f_L(\cdot)$ at $2x$ must be more than one-quarter its value at x. Hence, $f_L(x)$ cannot decrease too rapidly as x increases.

We next give three continuous distributions where Exchange Condition (8) is satisfied at least some of the time:

5. *Uniform distribution.* Let $f_L(x) = 1$, for $0 \leq x \leq 1$. Exchange Condition (8) is satisfied if, and only if, $x \leq 1/2$. This occurs with probability 3/4, because (8) fails exactly when $X > 1/2$; and $X > 1/2$ if, and only if, the independent events $X = L$ and $L > 1/2$ both occur, each of which has probability 1/2.

6. *Exponential distribution.* Let $f_L(x) = e^{-x}$, for $0 \leq x < \infty$. Exchange Condition (8) is satisfied if, and only if, $e^{-x} < 4e^{-2x}$, which is equivalent to $x < \ln 4 \cong 1.39$. It is

easy to verify that $\Pr(L \le \ln 4) = 3/4$ and $\Pr(S \le \ln 4) = 15/16$, so (8) is satisfied with probability

$$\Pr(X \le \ln 4) = (1/2)(3/4) + (1/2)(15/16) = 27/32 \cong .844.$$

7. *Exchange Condition* (8) *always satisfied.* Fix k so that $0 < k < 1$, and let $f_L(x) = (1-k)x^{-2+k}$ for $x \ge 1$ and $f_L(x) = 0$ for $0 < x < 1$. This function is a probability density because

$$\int_1^\infty x^{2+k} \, dx = \lim_{K \to \infty} \left[\frac{K^{-1+k}}{-1+k} - \frac{1^{-1+k}}{-1+k} \right] = \frac{1}{1-k}.$$

Then Exchange Condition (8) is satisfied for $x \ge 1$:

$$f_L(2x) = (1-k)(2x)^{-2+k} = 2^{-2}2^k(1-k)x^{-2+k} = \left(\frac{2^k}{4} \right) f_L(x) > \frac{f_L(x)}{4}.$$

Like example 3 in the discrete case, example 7 is a distribution with a floor: There is at least one observation (in this case, any x satisfying $1/2 \le x < 1$) for which you can be certain you have chosen the S box. Just as in the discrete case, it is possible to construct a "floorless" example in which Exchange Condition (8) is always satisfied. Again, the crucial feature is a probability density function $f_L(x)$ with a plateau, which falls off, but not too rapidly, as $x \to \infty$.

We now show

THEOREM 2. *For any probability density function $f_L(x)$, Exchange Condition* (8) *cannot fail for all values of x.*

Proof. Suppose, to the contrary, that

$$f_L(x/2) \ge 4f_L(x) \text{ for all } x \in (0, \infty). \tag{9}$$

Find an interval $[2^k, 2^{k+1}]$ such that

$$\int_{2^k}^{2^{k+1}} f_L(x) \, dx = K > 0. \tag{10}$$

Now (9) and (10) imply that

$$\int_{2^{k-1}}^{2^k} f_L(x) \, dx = \int_{2^k}^{2^{k+1}} f_L(y/2)(1/2) \, dy \ge 4(1/2) \int_{2^k}^{2^{k+1}} f_L(y) \, dy = 2K.$$

By induction, we can show that

$$\int_{2^{k-n-1}}^{2^{k-n}} f_L(x) \, dx \ge 2^n K$$

for $n = 0, 1, 2, \dots$. It follows that

$$\int_0^\infty f_L(x) \, dx \ge \sum_{n=0}^\infty \int_{2^{k-n}}^{2^{k-n+1}} f_L(x) \, dx \ge \sum_{n=0}^\infty 2^n K = \infty,$$

so $f_L(x)$ cannot be the probability density function of a continuous random variable.

Thus, no matter what continuous distribution L has, Exchange Condition (8) holds for some values of x, making switching profitable for those values. As example 7 demonstrates, switching may be profitable for all possible values of x.

5. Utilities

So far we have considered the contents of the boxes to be money. Moreover, we assumed that your objective is to maximize the expected amount of money you hold after your decision to switch or not. But the criterion of expected monetary values has been often criticized, going back at least to Bernoulli [2]. After all, if $x is a large prize, is $2x really worth twice as much to you?

The need to use expected values in game theory led von Neumann and Morgenstern [9] to develop a theory in which the subjective value of any outcome is measured by a real number called its *utility*, and rational decisions are those that maximize expected utility. The use of von Neumann-Morgenstern utilities is now standard in economics and elsewhere. For us, the box problem can also be expressed in utilities: Should you choose a gamble that either doubles or halves your utility with equal probability?

Our purpose in introducing utilities at this point is to highlight the relation of an important controversy in utility theory to the seemingly paradoxical situation in which you always prefer to switch boxes. We first show that if you always prefer to switch, the utility of the prize must be unbounded. Suppose otherwise. Then there is a number c such that the value, L, of the utility of the more valuable prize satisfies

$$\Pr(L > c) = 0 \text{ and } \Pr(c/2 < L \le c) > 0.$$

It follows that

$$\Pr(X = L | c/2 < X \le c) = 1,$$

and $\Pr(c/2 < X \le c) > 0$. Thus, General Exchange Condition (3) fails with positive probability. Therefore, if switching is always profitable, the (utility) value of the prize must be unbounded. As examples 2 and 6 demonstrate, however, this condition is necessary but not sufficient; utility may be unbounded, but for some observed values of the contents of B1, switching is not rational.

In fact, we can say more than this. Suppose that L, S, and X (now measured in utilities) are continuous and that Exchange Condition (8) holds for all $x \ge x_0$. Consider the expected value (assumed finite)

$$E = \int_{x_0}^{\infty} x f_L(x) \, dx = \int_{x_0}^{2x_0} x f_L(x) \, dx + \int_{2x_0}^{\infty} x f_L(x) \, dx.$$

Set

$$\int_{x_0}^{2x_0} x f_L(x) \, dx = H \ge 0,$$

and substitute $y = x/2$ in the second integral. This gives

$$E = H + \int_{x_0}^{\infty} 2 y f_L(2y) 2 \, dy = H + 4 \int_{x_0}^{\infty} y f_L(2y) \, dy > H + \int_{x_0}^{\infty} y f_L(y) \, dy \ge E.$$

This contradiction shows that the original integral must be divergent, and the (finite) expected value E cannot exist. In particular, if Exchange Condition (8) holds for all possible observations x, then the expectations of L, S, and X must be infinite.

A proof that expected utility must be infinite in the discrete case when Exchange Condition (6) is always satisfied is given by Nalebuff [6]. A related argument, suggested by Nyarko [9], goes as follows: If Y is the utility of the box not selected, and it is true for all x that

$$E(Y|X = x) > E(X|X = x),$$

then, by integration over all x,

$$E(Y) > E(X).$$

But, by symmetry, $E[Y] = E[X]$ because the two boxes have equal probabilities of containing either prize. Again, this contradiction shows that the unconditional expected values of the utilities can never exist.

What is the relevance of this calculation to the general properties of utility? In discussing the St. Petersburg paradox (see, e.g., [5]), Aumann [1] suggested that a distribution of prizes giving infinite expected utility cannot exist. But his position was disputed by Shapley [8], who claimed that the real problem was an empirical one—no rational gambler would believe that he or she would be paid arbitrarily large winnings.

In material terms, of course, there is never the possibility of unbounded support and infinite expected utility. Thus, following Aumann, you may have good reason to question the reality of examples 3, 4, and 7, in which switching is always profitable. But if you believe that utility for certain outcomes is immaterial and may be great beyond any measure, then it may be in your interest always to switch—given that these outcomes, according to your prior distribution, are sufficiently probable.

One might try to distinguish distributions with infinite expectation that always satisfy Exchange Conditions (6) and (8)—and make switching always rational—from those that do not. But such an attempt does not seem sensible, because (6) and (8) are "local" conditions that refer only to the probabilities associated with L at x and at $2x$, whereas infinite expected utility is a "global" condition that relates to the entire distribution of L.

In fact, a distribution may have infinite expected value, satisfy an Exchange Condition for arbitrarily large values, and yet fail this condition with positive probability, as shown by the following example. Let the density of L be

$$f_L(x) = \begin{cases} 0, & \text{if } x < 1 \\ 10^{-2k-1}, & \text{if } 10^k \le x < 10^{k+1} \text{ for } k = 0, 1, 2, \ldots \end{cases}$$

Thus, L has uniform density between 1 and 10, again between 10 and 100, etc. But at each power of 10, the density function drops to $1/100$ of its previous value. It is easy to verify directly that $f_L(x)$ is a density function, that $E(L) = \infty$, and that (8) holds (i.e., you should exchange) if $1/2 \le x < 1$ or, for some non-negative integer k, $10^k \le x < (5)10^k$. On the other hand, (8) fails (i.e., you should not exchange) if $(5)10^k \le x < 10^{k+1}$ for every $k = 0, 1, 2, \ldots$.

In this example, there is a positive probability of drawing a number that exceeds any preassigned bound. Within each "step" of the distribution of L, an exchange is called for if the number lies in the left-hand $4/9$, but not in the right-hand $5/9$, of the step. Thus, however large the preassigned bound is, there is a positive probability that exchanging a number exceeding it will be profitable, and a positive probability that it will not.

6. When Is Envy Rational?

Envy, according to Webster, is resentment caused by not possessing something you desire. If the desired object is truly unattainable, no gamble is worthwhile, so your envy cannot be attenuated. In the case of the box problem, however, we assume that you can transform your envy into a risky action to try to satisfy your desire.

We have shown that, for both discrete and continuous probability distributions, there is always some amount you can find in B1 that will provide grounds for envy. More surprising, for some distributions envy is always justified: *Every* amount in B1 is less than the expected value of the amount in B2, making a switch always profitable. However, this result depends on the distribution's having unbounded support and infinite expectation. That is, no matter what amount you find in B1, there is some possibility that the amount in B2 will be greater and that you will, in fact, receive a greater expected value from switching.

Like the St. Petersburg paradox, the box problem can be resolved by assuming that utility is bounded. Then it can never be the case that you will find it rational *always* to switch. On the other hand, if utility is unbounded and expected utility is infinite, always switching may be rational, depending on the prior distribution. So the paradox remains unresolved in this case.

The dependence of the "always switching" result on the distribution adds a new wrinkle to arguments about whether infinite expected utility is reasonable to posit. Even given the existence of infinite expected utility, we have shown that the shape of the distribution also matters: If it falls off sufficiently slowly, switching is always profitable. In this case, your envy will never be eliminated, no matter how much you find in B1.

The St. Petersburg paradox stimulated Bernoulli [2] and later mathematicians and economists to construct foundations on which to build a theory of utility and risk. Perhaps the box problem will stimulate the development of new foundations for viewing questions of exchange, including exchanges—unlike those discussed here—that are not simply one-person games against nature. Thus, in games in which two players must decide about an exchange, we have shown that the always-switching argument is sometimes reversed: No exchanges may be rational, no matter how small the amounts the players find in their boxes [3].

Acknowledgment. We thank the anonymous referees for their valuable comments on an earlier version of this paper.

REFERENCES

1. R. Aumann, The St. Petersburg paradox: A discussion of some recent comments *J. of Economic Theory* 14 (1977), 443–445.
2. D. Bernoulli, Exposition of a new theory on the measurement of risk (English translation of 1738 article), *Econometrica* 22 (1954), 23–36.
3. S. J. Brams, D. M. Kilgour, and M. D. Davis, Unraveling in games of sharing and exchange. In K. Binmore, A. Kirman, and P. Tani (eds.), *Frontiers of Game Theory*, MIT Press, Cambridge, MA, 1993, 195–212.
4. D. Gale, Mathematical entertainments, *Mathematical Intelligencer* 13 (1991), 31–32.
5. S. French, *Decision Theory: An Introduction to the Mathematics of Rationality*, Halsted/Wiley, New York, 1986.
6. B. Nalebuff, The other person's envelope is always greener, *J. of Economic Perspectives* 3 (1989), 171–181.
7. Y. Nyarko, Private communication to S. J. Brams (July 16, 1990).
8. L. S. Shapley, The St. Petersburg paradox: A con game? *J. of Economic Theory* 14 (1977), 439–442.
9. J. von Neumann and O. Morgenstern, *Theory of Games and Economic Behavior*, 3d edition, Princeton University Press, Princeton, NJ, 1953.

NOTES

Generating a Rotation Reduction
Perfect Hashing Function

ARLENE BLASIUS
State University of New York College at Old Westbury
Old Westbury, NY 11568

1. Introduction There are many applications where the retrieval of information from a table of identifier-information pairs is required. For example, consider a video rental store that uses the customers' phone numbers to identify their accounts. Using a phone number as an identifier, a clerk can access an account and retrieve such information as customer name, address, account balance, and so forth. All of the information associated with the phone number in this way is referred to as a *record*. In this paper, we are concerned with the mechanism that is used to obtain the record associated with a particular identifier, and in particular, with a process called *hashing*. In hashing, we visualize the records arranged in a table, the location of each record given by an address. A hashing function, h, is defined so that for each identifier w, $h(w)$ is an address in the table. Given the value of w, we compute $h(w)$ to obtain the address for w's record. In some applications, $h(w)$ is not exactly the right address for w, but is very close to the correct address. In these instances, we may examine the records near $h(w)$ until we locate the correct one. If we know in advance that $h(w)$ is exactly the right address, we say that it locates the desired record with a *single probe*.

In an ideal situation, all identity (ID) numbers would start at 0 and follow in sequence. Then we could use the ID numbers as the addresses, and take h to be the identity function. Unfortunately, in the real world this is not usually the case. In the example of the video store, the customer phone numbers would not be expected to occur in sequence. A table with one record for each possible phone number would be extremely sparse. Thus, although the retrieval procedure is simple and fast, the storage is used very inefficiently. On the other hand, we might simply store the phone numbers without any consideration to order, and search the list from the beginning each time a record is sought. This approach makes most efficient use of storage, but at the cost of execution speed. An optimal solution would be to find a function h that uses exactly as much storage as required for the data, and that returns a correct address with a single probe. Moreover, we would like the function to be defined in terms of simple arithmetic operations: addition, subtraction, multiplication and division (or reduction modulo a divisor).

As an example, let us consider the set of identifiers {49614, 50626, 49618, 53458, 49625, 54734, 54732, 54727, 50640, 50132, 53206, 50627}, and let us attempt to define a hashing function of the form $h(w) = w \bmod M$ for some M. This gives addresses in the range 0 to $M - 1$ so that at most M addresses are produced. To illustrate this idea, let us take M equal to 12. That is, let us reduce each identifier modulo 12, and attempt to use the result as the address of the associated record. This attempt is not always successful, for example $h(50626) = h(49618) = 10$. We describe this situation by saying that 50626 and 49618 *hash to the same position*, resulting in a *collision*.

Here is a simple algorithm for storing the data, using just 12 addresses, and resolving collisions. Consider the identifiers, in the order listed, calculating $h(w)$ for each. Assign the address of w to be $h(w)$, or if this produces a collision with a previous assignment, assign w the next available address. This produces the situation depicted in the Hash Table.

HASH TABLE

w	$h(w) = w \bmod 12$		
49614	6	0	53458
50626	10	1	54732
49618	10	2	54734
53458	10	3	50640
49625	5	4	53206
54734	2	5	49625
54732	0	6	49614
54727	7	7	54727
50640	0	8	50132
50132	8	9	50627
53206	10	10	50626
50627	11	11	49618

Note that as we construct the table, we can determine the maximum distance from $h(w)$ to the actual address used for w. For this table the maximum distance is 10, occurring when $w = 50627$. Therefore, to retrieve an identifier, we may have to probe as many as 11 times.

This example used a single operation, reduction modulo 12, for the hashing function. We might try other choices than 12 for the modulus, to see if we could reduce the number of probes. More generally, it is reasonable to expect that combining several arithmetic operations would produce a much richer set of functions from which to select h. Clearly, as we select a hashing function, our goal is to find a one-to-one function that minimizes the required storage space and computational difficulty. In the case of the previous example, we would like an arithmetic formula that provides a one-to-one map from the identifier set onto {0, 1, 2, 3, 4, 5, 6, 7, 8, 9, 10, 11}. This would give us a hashing function with perfect storage efficiency, and which locates any record with a single probe. Can you construct such a function?

This introductory section has discussed the use of hashing functions in designing a data storage and retrieval system. Note that in our approach the hashing function that is used dictates the order of storage of the data. This technique requires that we know at the outset the complete set of identifiers that will be used. There are situations where this is not possible. For instance, in the earlier example where customer phone numbers are the identifiers, we do not expect the designers of the retrieval system to know in advance what phone numbers will be entered into the system. For this paper, we restrict our attention to the case of a static identifier list, and construct hashing functions based on the assumption that we can order the data however we choose. The goal is to examine a class of simple functions that allows retrieval of an item from a minimal sized table with a single probe. Quite a number of other ways have been proposed to design hash functions. In a specific application one has to choose a suitable hash function comparing memory space required, execution time, and construction procedures. For further information about hashing functions, the reader should consult [3], [4]. [5], [8], [9], and [11].

2. Definitions   Let $\mathbb{N}$ be the set $\{0, 1, 2, 3, \ldots\}$ of nonnegative integers, $\mathbb{Z}$ be the set $\{\ldots, -1, 0, 1, \ldots\}$ of all integers, and $\mathbb{Z}_m$ be the set $\{0, 1, 2, \ldots, m - 1\}$. For any set A and function $f\colon A \to \mathbb{N}$, we let $f(A) = \{f(a)|a \in A\}$ so that

$$\min f(A) = \min\{f(a)|a \in A\} \quad \text{and} \quad \max f(A) = \max\{f(a)|a \in A\}.$$

The Euler ϕ function, $\phi(m)$, of an integer $m \neq 0$ is the number of positive integers less than or equal to $|m|$ and relatively prime to m. For example, $\phi(12) = |\{1, 5, 7, 11\}| = 4$.

Given a set $l = \{w_0, w_1, \ldots, w_{n-1}\} \subset \mathbb{N}$ we wish to construct a one-to-one hashing function $h\colon \mathbb{N} \to \mathbb{Z}_m$. Clearly, $m \geq n$. The best situation will occur when the length of the table is n; this assures that the table is full.

We now make the following definitions.

A *perfect hashing function* (phf for short) for l, is a one-to-one function h mapping l to $\mathbb{Z}_m$ for some m such that $\min h(l) = 0$. A *minimal* perfect hashing function for l, $|l| = n$, satisfies $m = n$. The *load factor*, α, of a phf is n/m, the ratio of the number n of occupied entries in the table to the length of the table.

Given a finite set $l \subset \mathbb{N}$ we wish to choose integers q, s, M and N so that the *remainder reduction hashing function*, namely, function

$$h(w) = \lfloor((qw + s) \bmod M)/N\rfloor$$

is a phf. We would like to choose the constants so that h is perfect, and so that α is as close to unity as possible. As a first step, choose any $M \notin \{d \in \mathbb{N}|d \text{ is a divisor of } u - w \text{ for some } u, w \in l\}$. This guarantees that $w \not\equiv u \pmod M$ for any pair $u, w \in l$.

The next question is how to pick N and s so that $\min h(l) = 0$ and $m = \max h(l) + 1$ is minimized. To simplify the notation, we assume that the elements of l have been reduced modulo M and indexed in ascending order forming the set l_M, $l_M \subset \mathbb{Z}_M$. We retain the subscript on l to remind us of the value of M.

To choose N and s, we now consider the *scrambled* set, $l_M = \{w_i \bmod M|w_i \in l\}$ $\subset \mathbb{Z}_M$, and examine the simpler function

$$h(w) = \lfloor((w + s) \bmod M)/N\rfloor,$$

which we call a *rotation reduction*, perfect hashing function. We call s the *rotation value*. Clearly, adding s to each w and reducing modulo M results in a 'rotation' of the ordering of the w's in l_M. The set l_M will be called *rotationally reducible* by $N \in \mathbb{Z}$ if, and only if, there exists a rotation value $s \in \mathbb{Z}$ such that $h(w) = \lfloor((w + s) \bmod M)/N\rfloor$ is a rotation reduction phf for l_M.

Example 1. Given the set $l_{17} = \{4, 6, 7, 12, 14, 16\}$ and $N = 3$, we look successively at all possible s values, $0 \leq s < 17$, and find that l_{17} is rotationally reducible by $N = 3$. The function $h(w) = \lfloor((w + 10) \bmod 17)/3\rfloor$ is a minimal rotation reduction phf for l_{17}.

w	4	6	7	12	14	16
$(w + 10)\bmod 17$	14	16	0	5	7	9
$h(w)$	4	5	0	1	2	3

3. The rotation reduction phf   In this section we develop a criterion for rotational reducibility. Throughout, we use the notation $l_M = \{w_0, w_1, \ldots, w_{n-1}\} \subset \mathbb{Z}_M$, where the elements w_i are assumed to appear in ascending order. We define $\{\delta_0, \delta_1, \ldots, \delta_{n-1}\}$

by the set of first differences thus: $\delta_i = w_{i+1} - w_i$, $0 \le 1 < n - 1$ and $\delta_{n-1} = M + w_0 - w_{n-1}$. Observe that each $\delta_i \in \mathbb{Z}_M$. As a further notational convience, we agree that all subscript addition is done modulo n; for example $w_{(n-1)+1} = w_0$.

THEOREM 1. *For a given integer u, let s_u be the unique index i such that $(w_{i+1} + u) \bmod M < (w_i + u) \bmod M$. Then l_M is rotationally reducible by N if, and only if, there exists u such that $\delta_i + ((w_i + u) \bmod M) \bmod N \ge N$ for all $i \ne s_u$.*

Proof. We assume that the w's are in ascending order. When we add $u \in \mathbb{Z}$ to each w and reduce modulo M, the order gets shifted. Thus, if $v_i = (w_i + u) \bmod M$, then the v's are ordered like this: $v_{j+1}, v_{j+2}, \ldots, v_{n-1}, v_0, \ldots, v_j$ (for some j). Clearly, $v_{i+1} - v_i = \delta_i$ for all $i \ne j$; and $v_{j+1} - v_j = \delta_j - M$. Now $v_{j+1}, v_{j+2}, \ldots, v_{n-1}, v_0, \ldots, v_j$ is an increasing sequence in $\mathbb{Z}_M$, so integer division by N results in a nondecreasing sequence. To be sure that h is an injection, we need only guarantee that the sequence is strictly increasing. Letting $v_i = Nq + r$, $0 \le r < N$, we have $v_{i+1} = \delta_i + v_i = Nq + \delta_i + r$. Thus h is an injection if, and only if, the remainder of the division v_i/N plus the gap δ_i is at least N. Assuming that such a u exists, $h'(w) = \lfloor ((w + u) \bmod M)/N \rfloor$ is an injective mapping from l_M into $\mathbb{Z}_M$. To insure that the minimum value is 0 we now look at $h'(w) - \min h'(l_M) = \lfloor \{(w + u) \bmod M - N(\min h'(l_M))\}/N \rfloor$. Since $0 \le N (\min h'(l_M)) < M$ and $w + u \ge N(\min h'(l_M))$ we have $(w + u) \bmod M - N (\min h'(l_M)) = \{w + u - N(\min h'(l_M))\} \bmod M$. Therefore, $h(w) = \lfloor ((w - s) \bmod M)/N \rfloor$, where $s = u - N(\min h'(l_M))$, is a rotation reduction phf for l_M.

To simplify notation, we now let $d_i = (w_{i+1} + u) \bmod M - (w_i + u) \bmod M$.

COROLLARY 1. *Each $\delta_i \in \mathbb{Z}_M$ (as noted earlier).*

COROLLARY 2. *$-M + 1 \le d_i \le M - 1$ and $\delta_i = d_i \bmod M$.*

COROLLARY 3. *d_i equals either δ_i or $\delta_i - M$, and the latter case obviously only occurs when d_i is negative.*

Example 2.

a.

$l_{23} =$	{1,	2,	4,	6,	14,	15,	19,	21} ; $u = 10$
δ_i	1	2	2	8	1	4	2	3
$(w_i + u) \bmod 23$	11	12	14	16	1	2	6	8
d_i	1	2	2	−15	1	4	2	3

b.

$l_{18} =$	{0,	3,	5,	8,	9,	10,	12,	14} ; $u = 0$
δ_i	3	2	3	1	1	2	2	4
$(w_i + u) \bmod 18$	0	3	5	8	9	10	12	14
d_i	3	2	3	1	1	2	2	−14

Thus we see that the last row of each table is the same as the second row, with *one* exception as predicted by the corollary. Hence we look for a u such that $u \in \bigcap_{0 \le i < n} J(w_i)$, where

$$J(w_i) = \{u | \delta_i + ((w_i + u) \bmod M) \bmod N \ge w \quad \text{or} \quad (w_{i+1} + u) \bmod M < (w_i + u) \bmod M\}.$$

Obviously, if $\delta_i \ge N$, $J(w_i) = \mathbb{Z}_M$ and no computation is necessary, l_M is rotationally reducible by N if, and only if, $J = \bigcap_{0 < i < n} J(w_i)$ is nonempty.

Example 3. To see that $l_{17} = \{1, 6, 7, 12, 14, 16\}$ is rotationally reducible by 3, consider

$$J(1) = J(7) = Z_{17}$$
$$J(6) = \{12, 5, 8, 10, 13, 16\}$$
$$J(12) = \{1, 2, 3, 4, 6, 7, 9, 10, 12, 13, 15, 16\}$$
$$J(14) = \{0, 1, 2, 4, 5, 7, 8, 10, 11, 13, 14\}$$
$$J(16) = \{0, 2, 3, 5, 6, 8, 9, 11, 12, 14, 15, 16\}.$$

Since $J = \{2\}$, l_{17} is rotationally reducible by 3 using $u = 2$ as the rotation value; the hashing function would be $h(w) = \lfloor (w + 2) \bmod 17)/3 \rfloor$.

When $N = 2$, Theorem 1 simplifies to the following. (Parity refers to the property of being even or odd.)

COROLLARY 1'. *l_M is rotationally reducible by $N = 2$ if, and only if, there exists a u such that for all $w_i \in l_M$ with $\delta_i = 1$, $(w_i + u) \bmod M$ has odd parity or $(w_i + u) \bmod M = M - 1$.*

COROLLARY 2'. *If M is even, then l_M is rotationally reducible by $N = 2$ if, and only if, all the w_i;s with $\delta_i = 1$ have the same parity P.*

(Choose u with parity P', $P' \neq P$, then $h(w) = \lfloor ((w + u) \bmod M)/2 \rfloor$ will be injective.)

COROLLARY 3'. *If M is odd, then l_M is rotationally reducible by $N = 2$ if, and only if, there exists a $k \in [0, n - 1]$ such that for all $i < k$ the w_i's with $\delta_i = 1$ have the same parity P, and for all $i > k$ the w_i's with $\delta_i = 1$ have the same parity P', with $P \neq P'$.*

(Choose k such that w_k has parity P' and let $u = M - 1 - w_k$, then $h(w) = \lfloor ((w + u) \bmod M)/2 \rfloor$ will be injective.)

The preceding results help us search for a possible rotation reduction perfect hashing function. For some l_M and N, there may be no such function. However, if $q \in G_M$, where G_M is the set of integers q such that $0 < q \leq |M|$ and q is relatively prime to m, the set

$$ql_M = \{qw \bmod M | w \in l_M\}$$

is a subset of Z_M with n distinct elements because q is prime to M. In general, if $q, q' \in G_M$ and $q \neq q'$, the set ql_M and $q'l_M$ may be different, and we have the possibility of investigating many sets ql_M for a rotation reduction phf. By the following theorem, due to Sprugnoli [8, p. 847], at most $\phi(M)/2$ sets ql_M need be investigated.

THEOREM 2. *The set ql_M is rotationally reducible by N if, and only if, $(M - q)l_M$ is also.*

We provide an outline of the proof, using an alternate method to that of Sprugnoli. In the interest of brevity, we omit the details.

Letting $h(x) = \lfloor ((x + s) \bmod M)/N \rfloor$ be a rotation reduction phf for ql_M, one can show that $h'(x) = \lfloor ((x + u) \bmod M)/N \rfloor$, where $u = N (\max h(ql_M)) - s - 1$, is an injective mapping from $(M - q)l_M$ into $\mathbb{N}$. If $\min h'((M - q)l_M) \neq 0$ then replace u with $u - N (\min h'((M - q)l_M))$, and it follows that u determines a rotation reduction phf for $(M - q)l_M$.

Example 4. Letting $l_{29} = \{0, 4, 8, 9, 16, 17, 18, 22, 28\}$ the set $10l_{29} = \{0, 3, 6, 11, 15, 17, 19, 22, 25\}$ is rotationally reducible by $N = 3$. A rotation reduction phf for $10l_{29}$ is $h(x) = \lfloor ((x + 19) \bmod 29)/3 \rfloor$ and the corresponding rotation reduction phf for $19l_{29} = \{0, 4, 7, 10, 12, 14, 18, 23, 26\}$ is $h(x) = \lfloor((x + 4) \bmod 29)/3\rfloor$. We note that the rotation reduction phf for $10l_{29}$ is minimal and the one for $19l_{29}$ is not; the load factors are 1 and $9/10$, respectively.

4. The remainder reduction phf

We now turn our attention to remainder reduction phf's:

$$h(w) = \lfloor((qw + s) \bmod M)/N\rfloor.$$

In order that the elements in l can be mapped onto different values, it must be that $N(n - l) < M$. Thus, for any $M \notin \{d \in \mathbb{N} \mid d$ is a divisor of $u - w$ for some $u, w \in l\}$, there is only a finite number of values q, s, N corresponding to possible remainder reduction phf's for l. Thus a procedure can be devised to perform an exhaustive search for such parameters.

5. Some comments and unsolved questions

Given the original set l, we can always take M to be $w_{n-1} - w_0 + 1$, take s so that the least element gets shifted to 0, and then take $N = 1$. That is, the trivial hashing function $h(w) = w - w_0$ is a remainder reduction phf. The efficiency is n/M. We have not been able to prove that given a bound on $\alpha > n/(w_{n-1} - w_0 + 1)$ a remainder reduction phf always exists. Nor is anything known about how to calculate the remainder reduction phf with maximal efficiency for an arbitrary set l. By taking sufficiently large M is it possible to obtain an almost minimal phf? Is there an example where for a particular l there is a maximal possible efficiency < 1 for remainder reduction phf's? We cannot predict what the best possible efficiency is for an arbitrary set l.

Using the methods above we have been able to calculate a minimal phf for the set considered in the introduction, $\{49614, 50626, 49618, 53458, 49625, 54734, 54732, 54727, 50640, 50132, 53206, 50627\}$; $h(w) = \lfloor((21w + 92) \bmod 101)/9\rfloor$. The Hash Table is given below.

HASH TABLE

	ID
0	50640
1	50626
2	54734
3	50627
4	50132
5	49618
6	53206
7	49614
8	54727
9	54732
10	53458
11	49625

ACKNOWLEDGEMENTS. It is a pleasure to thank David Lubell for helpful suggestions and conversations. We thank the Mathematics and Computer Science Department of Adelphi University, and the Computing Centers of Adelphi University and SUNY College at Old Westbury for providing the support and facilities needed for the development of the paper. The author also thanks the referees for their generous suggestions, which have significantly improved the presentation.

REFERENCES AND RELATED READINGS

1. A. V. Aho, J. E. Hopcroft, and J. D. Ullman, *Data structures and Algorithms*, Addison-Wesley Publishing Co., Reading, MA, 1983.
2. M. R. Anderson, M. G. Anderson, Comments on perfect hashing functions: A single probe retrieving method for static sets, *Comm. ACM* 22 (1979), 104.
3. R. J. Cichelli, Minimal perfect hash functions made simple, *Comm. ACM* 23 (1980), 17–19.
4. G. V. Cormack, R. N. S. Horspool, and M. Kaiserworth, Practical perfect hashing, *Computer J.* 28 (1985), 54–58.
5. G. Jaeschke, Reciprocal hashing: A method for generating minimal perfect hashing functions, *Comm. ACM* 24 (1981), 829–833.
6. D. E. Knuth, *The Art of Computer Programming*, Vol. 1–2. Addison-Wesley Publishing Co., Reading, MA, 1975.
7. H. F. Smith, *Data Structures, Form and Function*, Harcourt Brace Jovanovich, New York, 1987.
8. R. Sprugnoli, Perfect hashing functions: A single probe retrieving method for static sets, *Comm. ACM* 20 (1977), 841–850.
9. T. J. Sager, A polynomial time generator for minimal perfect hash functions, *Comm. ACM* 28 (1985), 523–532.
10. J. V. Uspensky, and M. A. Heaslet, *Elementary Number Theory*, McGraw-Hill, New York, 1939.
11. W. P. Yang, and M. W Du, A backtracking method for constructing perfect hashing functions from a set of mapping functions, *BIT* 25 (1985), 148–164.

Proof without Words:
On a Property of the Sequence of Odd Integers (Galileo, 1615)

$$\frac{1}{3} = \frac{1+3}{5+7} = \frac{1+3+5}{7+9+11} = \cdots.$$

$$\frac{1+3+\cdots+(2n-1)}{(2n+1)+(2n+3)+\cdots+(4n-1)} = \frac{1}{3}$$

REFERENCE

1. S. Drake, *Galileo Studies*, The University of Michigan Press, Ann Arbor (1970), pp. 218–219.

—Roger B. Nelsen
Lewis and Clark College
Portland, OR 97219

The Birth of Period Three

PARTHA SAHA
STEVEN H. STROGATZ
The Massachusetts Institute of Technology
Cambridge, MA 02139

1. The logistic map   The logistic map is one of the most far-reaching examples in all of mathematics [1–8]. It is given by the difference equation

$$x_{n+1} = rx_n(1 - x_n), \tag{1}$$

where $0 \le x_n \le 1$ and $0 \le r \le 4$. In other words, given some starting number $0 \le x_1 \le 1$, we generate a new number x_2 by the rule $x_2 = rx_1(1 - x_1)$, and then repeat the process to generate x_3 from x_2, and so on.

This equation has many virtues:

1) It is *accessible*. High school students can explore its patterns, as long as they have access to a hand calculator or a small computer.

2) It is *exemplary*. This single example illustrates many of the fundamental notions of nonlinear dynamics, such as equilibrium, stability, periodicity, chaos, bifurcations, and fractals. May [6] was the first to stress the pedagogical value of (1).

3) It is *living mathematics*. Most of the important discoveries about the logistic map are less than 20 years old. Certain aspects of (1) are still not understood rigorously, and are being pursued by a few of the finest living mathematicians.

4) It is *relevant to science*. Predictions derived from the logistic map have been verified in experiments on weakly turbulent fluids, oscillating chemical reactions, nonlinear electronic circuits, and a variety of other systems [8].

2. Period-3 cycle and tangent bifurcation   This paper is concerned with one aspect of the logistic map, namely the value of r at which a period-3 cycle is created in a tangent bifurcation. To explain what this mouthful means (and why anyone might care!), we begin with an example. If we set $r = 3.835$ and then generate the sequence $\{x_n\}$, we find that x_n eventually repeats every three iterations. This is shown graphically in FIGURE 1. For a typical choice of x_1, the sequence bobbles around for a few iterations and finally approaches a *period-3 cycle*, for which $x_{n+3} = x_n$.

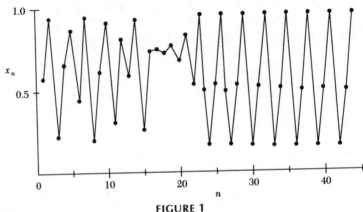

FIGURE 1
Time series of x_n, for $r = 3.835$.

Of course, the parameter value $r = 3.835$ was cunningly chosen; for different choices of r, one can see completely different long-term behavior of x_n. To see the behavior for all values of r at the same time, we plot the well-known "orbit diagram" [2, 3] for the system (FIGURE 2). This picture should be regarded as a stack of vertical lines, one above each r. For a given r, we start at some random x_1, and then iterate for 300 cycles or so, to allow the system to settle down to its eventual behavior. Now that the transients have presumably decayed, we plot many points, say $x_{301}, \ldots, x_{600}$ above that r. Then we move on to the next r and repeat, eventually sweeping across the whole picture.

FIGURE 2 shows the most interesting part of the diagram, in the region $3.4 \le r \le 4$. At $r = 3.4$, the system exhibits a period-2 cycle, as indicated by the two branches. As r increases, these branches split, yielding a period-4 cycle. This splitting is called a "period-doubling bifurcation". A cascade of further period-doublings occurs as r increases, yielding period-8, period-16, \ldots, until at $r \approx 3.57$ the map becomes "chaotic". The orbit diagram seems to have degenerated into a featureless mass of dots. Yet order sometimes re-emerges from chaos as we move to still larger r. This is seen most dramatically in the period-3 window marked on FIGURE 2. This region includes the value $r = 3.835$ used earlier in FIGURE 1.

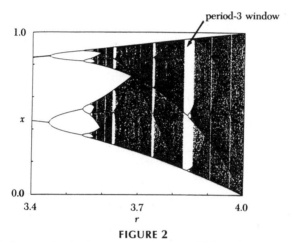

FIGURE 2
Orbit diagram for the logistic map. (From ref. [1], with permission.)

To understand how the period-3 cycle is born from chaos, we first need to introduce some notation. Let $f(x) = rx(1 - x)$ so that (1) becomes $x_{n+1} = f(x_n)$. Then $x_{n+2} = f(f(x_n))$ or more simply, $x_{n+2} = f^2(x_n)$. Similarly, $x_{n+3} = f^3(x_n)$. The function $f^3(x)$ is the key to understanding the birth of the period-3 cycle. Any point p in a period-3 cycle repeats every three iterates by definition, so such points satisfy $p = f^3(p)$. Since $f^3(x)$ is an eighth-degree polynomial, this equation is not explicitly solvable. But a graph provides sufficient insight. FIGURE 3 plots $f^3(x)$ for $r = 3.835$. Intersections between the graph and the diagonal line correspond to solutions of $f^3(x) = x$. There are eight solutions, six of interest to us and marked with dots, and two imposters that are not genuine period-3; they are actually fixed points, i.e. period-1 points for which $f(x^*) = x^*$. The black dots in FIGURE 3 correspond to the stable period-3 cycle seen in FIGURE 1, whereas the open dots correspond to an unstable cycle that is not observed numerically.

Now suppose we decrease r into the chaotic regime—how does the graph change? FIGURE 4 shows that when $r = 3.8$, the six marked intersections have vanished.

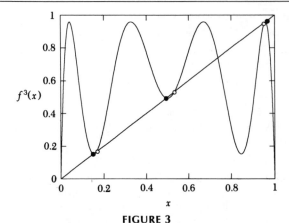

FIGURE 3

Graph of $f^3(x)$, for $r = 3.835$. Black dots, stable period-3 points; open dots, unstable period-3 points.

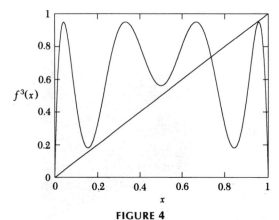

FIGURE 4

Graph of $f^3(x)$ for $r = 3.8$. The period-3 cycle has disappeared.

Somewhere between $r = 3.8$ and $r = 3.835$, the graph of $f^3(x)$ must have become tangent to the diagonal. At this critical value of r, the period-3 cycles are created in a *tangent bifurcation*.

Finally we come to the point of this paper. In several texts on chaos, it is mentioned that the value of r at the tangent bifurcation is given exactly by $1 + 2\sqrt{2} \approx 3.8284\ldots$ (e.g. [5, p. 169], [7, p. 289], [8, p. 83]). Given the beautiful simplicity of this result, one of us (Strogatz) assumed it should be easy to derive, and assigned it as a homework problem in a class on nonlinear dynamics. Of course, there were grounds for suspicion: The result is always stated without proof in the references we have seen. A few students in the class managed to derive the result with the help of Maple, MACSYMA or Mathematica—but these solutions were unsatisfying. One student (Saha) found an elementary solution that we present here. The solution exploits the symmetries of the period-3 cycle, and illustrates the importance of finding the right change of variables.

3. The period-3 conditions The period-3 conditions can be expressed in terms of the three points x, y, z in the cycle:

$$y = rx(1 - x) = f(x), \tag{2}$$

$$z = ry(1 - y) = f(y) = f^2(x), \tag{3}$$

$$x = rz(1 - z) = f(z) = f^2(y) = f^3(x). \tag{4}$$

We are also given that the onset of period-three is heralded by a tangent bifurcation. Hence f^3 has slope 1 at each intersection with the diagonal. At x, this yields

$$\frac{d(f^3(x))}{dx} = \frac{d(f^3(x))}{d(f^2(x))} \cdot \frac{d(f^2(x))}{d(f(x))} \cdot \frac{d(f(x))}{dx}$$

$$= \frac{d(f(z))}{dz} \cdot \frac{d(f(y))}{dy} \cdot \frac{d(f(x))}{dx}$$

$$= r^3(1 - 2z)(1 - 2y)(1 - 2x)$$

$$= 1. \tag{5}$$

Equations (2)–(5) are four equations in four unknowns: x, y, z, r. Can we solve for r analytically? The answer is yes, though straightforward attempts (like collapsing the four equations into two in x and r) quickly get out of hand. The system suggests changes of variables that considerably simplify the process. We show how.

4. Two smaller problems Two subsequent changes of variables break the problem into two easily manageable ones. We first notice that the right-hand sides of (2)–(4) suggest a certain symmetry of the three variables x, y, and z about the value $1/2$. Accordingly, we define the variables $p = x - \frac{1}{2}$, $q = y - \frac{1}{2}$, and $t = z - \frac{1}{2}$. Another change of variables, $A = rp$, $B = rq$, $C = rt$, renders (2)–(4) very simple:

$$\frac{r^2}{4} - \frac{r}{2} = A^2 + B = B^2 + C = C^2 + A. \tag{6}$$

Equation (5) is even simpler:

$$8ABC = -1. \tag{7}$$

We thus have two smaller problems to solve. If we let R denote the common value of the terms in (6), we get

$$R = A^2 + B = B^2 + C = C^2 + A, \tag{8}$$

and a quadratic equation in r,

$$\frac{r^2}{4} - \frac{r}{2} = R. \tag{9}$$

Our strategy is to solve (7) and (8) for R, and then invoke (9) to obtain r.

If we now try to find the values for A, B, and C, we will run into an avoidable complication. We must first realize that period three, by definition, implies three distinct values of A, B, and C. Looking at (8), we notice that cyclic interchanges in A, B, and C (i.e. $A \to B \to C \to A$) leave it unchanged. Thus if we solved for the variables A, B, and C, we would be forced to find not only the different triplets of numbers allowed *but their cyclic reassignments as well!* This clutter of solutions can clearly be reduced by another change of variables. We realize that each triplet and all its cyclic variations satisfy a *single cubic equation* of the form, $(x - A)(x - B)(x - C) = 0$. The coefficients of the cubic equation are independent of the cyclic reassignments, and are thus the variables we should use. Expanding and collecting powers of x of this cubic, we see that the coefficients are:

$$a = A + B + C, \qquad (10)$$

$$b = AB + BC + CA, \qquad (11)$$

$$c = ABC. \qquad (12)$$

Our ultimate task is to solve an equation in one of the new variables; for reasons that become clear later, we will choose a. Equality in any of the two variables among A, B, and C would imply equality in all three, as can be seen from (8). This would result in $8A^3 = -1$, which (since $a = 3A$ in this case) becomes $8a^3 + 27 = 0$ or

$$(2a + 3)(4a^2 - 6a + 9) = 0. \qquad (13)$$

What happens when $A = B = C$? We get the period-1 condition mentioned in Section 2. Thus, when we finally derive an equation for a, we expect it to contain these factors, corresponding to the period-1 "imposters" mentioned earlier. We will be interested in any *additional* factors that may be present, as they will be related to the genuine period-3 solutions.

5. Manipulations For much of this section, we will be manipulating algebraic expressions. There is really nothing conceptual involved—just lots of *careful* algebraic maneuvers. The goal is to derive an equation in one of the variables among a, b, and c.

We will need the following algebraic identities:

$$A^2 + B^2 + C^2 = a^2 - 2b, \qquad (14)$$

$$A^3 + B^3 + C^3 = a^3 - 3ab + 3c, \qquad (15)$$

$$(AB)^2 + (BC)^2 + (CA)^2 = b^2 - 2ca. \qquad (16)$$

Adding the three expressions for R in (8) and using the identities (14)–(16), we obtain

$$R = \tfrac{1}{3}(a^2 + a - 2b). \qquad (17)$$

Now multiplying the first expression for R in (8) by A, the second by B, and the third by C, adding and using (14)–(17), we get

$$2a^3 - 7ab - a^2 + 9c + 3b = 0. \qquad (18)$$

Next we multiply the first expression for R by C, the second by A, and the third by B, add and compare the resulting equation with what we had after a similar process just before—this establishes

$$A^3 + B^3 + C^3 = A^2C + B^2A + C^2B = a^3 - 3ab + 3c. \qquad (19)$$

To obtain the next equation, we combine two of the three expressions for R as

$$(A^2 + B)(B^2 + C) = R^2, \quad \text{etc.} \qquad (20)$$

In a manner we have followed before, we add the above three equations, and then use (14)–(17) and (19) to obtain:

$$a^4 - 4a^3 + 14ab + a^2 + b^2 + 6ac - 4a^2b - 3b - 18c = 0. \qquad (21)$$

6. Solving for r Now it's time to reap the fruits of our labors. Equation (7) gives us $c = -1/8$ which, when put in (18), lets us solve for b:

$$b = \frac{16a^3 - 8a^2 - 9}{56a - 24}. \tag{22}$$

Substituting the above b in (21), we get the long-awaited equation in a:

$$1536a^6 - 3072a^5 + 4608a^4 + 3456a^3 - 10368a^2 + 15552a - 5832 = 0. \tag{23}$$

The above factors into:

$$24(2a - 1)(2a + 3)(4a^2 - 6a + 9)^2 = 0. \tag{24}$$

The factors $2a + 3$ and $4a^2 - 6a + 9$ were expected from (13). The only value of interest for a is then $1/2$ and thus, from (22), $b = -9/4$. Using (17), we finally obtain $R = 7/4$ and now going back to the quadratic equation in r (9), we see that the only nonnegative root is $r = 1 + 2\sqrt{2}$, as desired.

7. Generalizations After this paper was completed, we discovered that similar methods have been used by Hitzl and Zele [4] in their study of the Henon map, a two-dimensional generalization of the logistic map. They find conditions for the existence of periods 1–6. Their analysis could be used as a challenging follow-up to the simpler case considered here.

Acknowledgements. We thank Bob Devaney, John Guckenheimer, Denny Gulick, and E. Atlee Jackson for helpful discussions, and Harvey Kaplan for bringing the paper of Hitzl and Zele to our attention. S. H. S. acknowledges support from an NSF Presidential Young Investigator Grant.

REFERENCES

1. D. Campbell, in *Lectures in the Sciences of Complexity*, D. L. Stein, ed., Addison-Wesley Advanced Book Program, Redwood City, CA, 1989.
2. R. L. Devaney, *An Introduction to Chaotic Dynamical Systems*, 2nd edition, Addison-Wesley Advanced Book Program, Redwood City, CA, 1989.
3. D. Gulick, *Encounters with Chaos*, McGraw-Hill, San Francisco, 1992.
4. D. L. Hitzl and F. Zele, An exploration of the Henon quadratic map, *Physica* D, 14 (1985), pp. 305–326.
5. E. A. Jackson, *Perspectives of Nonlinear Dynamics*, Vol. 1, Cambridge University Press, Cambridge, UK, 1989.
6. R. M. May, Simple mathematical models with very complicated dynamics, *Nature*, 261 (1976), pp. 459–467.
7. M. Schroeder, *Fractals, Chaos, Power Laws*, W. H. Freeman, New York, 1991.
8. H. G. Schuster, *Deterministic Chaos: An Introduction*, 2nd edition, VCH, Weinheim, West Germany, 1989.

A Combinatorial Proof of the Pythagorean Theorem

EDWARD R. SCHEINERMAN*
The Johns Hopkins University
Baltimore, MD 21218-2689

Can the quintessential theorem of plane geometry—the Pythagorean—be understood and proved as a theorem about finite sets? Yes! Here's how.

There are two steps that take us from the Pythagorean theorem to finite sets. First, we rewrite the Pythagorean theorem as the trigonometric identity $\sin^2 x + \cos^2 x = 1$. Second, we see this identity as an equation involving exponential generating functions. The first step is simple; let us focus on exponential generating functions (see [1] for an excellent discussion of generating functions). Given a sequence of numbers, $a_0, a_1, a_2, \ldots$, its *exponential generating function* is

$$A(x) = \sum_{n=0}^{\infty} a_n \frac{x^n}{n!}.$$

We need only one fact about exponential generating functions, namely, the multiplication rule. Suppose $A(x)$, $B(x)$, and $C(x)$ are the exponential generating functions for sequences $\{a_n\}$, $\{b_n\}$, and $\{c_n\}$, respectively. If $C(x) = A(x)B(x)$, how does c_n relate to the a's and the b's? The answer is the following convolution formula that the reader can easily check:

$$c_n = \sum_{k=0}^{n} \binom{n}{k} a_k b_{n-k}.$$

Thus if we write

$$\sin x = x - \frac{x^3}{3!} + \frac{x^5}{5!} - \frac{x^7}{7!} + \cdots = \sum_{n=0}^{\infty} s_n \frac{x^n}{n!},$$

$$\cos x = 1 - \frac{x^2}{2!} + \frac{x^4}{4!} - \frac{x^6}{6!} + \cdots = \sum_{n=0}^{\infty} c_n \frac{x^n}{n!},$$

the Pythagorean theorem ($\sin^2 x + \cos^2 x = 1$) can be written:

$$\sum_{k=0}^{n} \binom{n}{k} s_k s_{n-k} + \sum_{k=0}^{n} \binom{n}{k} c_k c_{n-k} = \begin{cases} 1 & n=0 \\ 0 & n>0 \end{cases}. \tag{1}$$

When $n = 0$ the only nonzero term is $\binom{0}{0} c_0 c_0 = 1$, so let us focus on the case $n > 0$. If n is odd, all the terms $s_k s_{n-k}$ and $c_k c_{n-k}$ are zero, so the statement is trivial. However, when n is even, we check that (1) reduces to

$$\sum_{k=0}^{n} (-1)^k \binom{n}{k} = 0. \tag{2}$$

Now (2) is an immediate consequence of the binomial theorem, but we promised the reader an interpretation about finite sets. Separating positive and negative terms,

---

*Research supported in part by the Office of Naval Research.

we have

$$\sum_{k \text{ odd}} \binom{n}{k} = \sum_{k \text{ even}} \binom{n}{k} \tag{3}$$

that says: The number of odd cardinality subsets on an n-set equals the number of subsets of even cardinality. To finish, we need a one-to-one correspondence between the odd and even subsets of an n-set. When n is odd, taking complements does the trick. But our job is to prove (2) (or (3)) for n even! This is almost as easy. Let $N = \{1, 2, \ldots, n\}$. Then for any $A \subseteq N$, let

$$A' = \begin{cases} A \cup \{1\} & \text{when } 1 \notin A, \text{ and} \\ A \setminus \{1\} & \text{when } 1 \in A. \end{cases}$$

Finally, observe that $A \leftrightarrow A'$ gives a one-to-one pairing of the even and the odd subsets of N.

Cute, but is it a proof? Yes. One "only" needs to check that the Taylor series for $\sin x$ and $\cos x$ are correct and bring some real analysis to bear. To this end, we need to know that $\frac{d}{dx} \sin x = \cos x$, $\frac{d}{dx} \cos x = -\sin x$, $\sin 0 = 0$ and $\cos 0 = 1$. The derivatives can be derived from standard formulas such as $\sin(x + y) = \sin x \cos y + \sin y \cos x$ and basic limits such as $\frac{\sin x}{x} \to 1$ as $x \to 0$. These can be verified using geometric arguments that do not require the use of the Pythagorean Theorem.

REFERENCE

1. Herbert S. Wilf, *Generatingfunctionology*, Academic Press, 1990.

Math Bite: Normality of the Commutator Subgroup

The commutator subgroup C of a group G is the smallest subgroup of G containing all elements of the form $aba^{-1}b^{-1}$, where a and b are arbitrary elements of G. To see that C is a normal subgroup of G, let c be a member of C, and let g be any element of G. Note that $gcg^{-1}c^{-1}$ is in C, whence by closure $gcg^{-1}c^{-1}c = gcg^{-1}$ must also belong to C.

—LAWRENCE MYERS
RESEARCH TRIANGLE INSTITUTE
RESEARCH TRIANGLE PARK, NC 27709

Whiskey, Marbles, and Potholes

J. CHRIS FISHER
DENIS HANSON
University of Regina
Regina, Saskatchewan, Canada S4S 0A2

The 1990 TV movie *Gunsmoke: The Last Apache*, which resurrected the Matt Dillon role for James Arness, added an intriguing twist to the traditional duel: Six identical shot glasses of whiskey were set on the bar, of which one was generously laced with deadly strychnine. The good guy drank first, survived, then laughed triumphantly, "Ha, you should have gone first. The first man to drink had a numerical advantage!" **Question:** Was he laughing because he thought the scriptwriters so ridiculously inept at mathematics?

Here is a pair of problems that are well suited for an elementary course in finite mathematics. The first is routine; the second, a simple variation of the first, has a somewhat surprising answer.

PROBLEM A. *Two players take turns drawing a single marble from an urn containing one red marble and four that are drab. They continue to remove marbles until one player wins by drawing the red marble. For the game to be fair, how much should player I (who picks first) pay when player II puts up $2?* (Answer: $3)

PROBLEM B. *Player II makes an offer to his opponent: For 50¢ he will increase the probability of an immediate victory by removing one of the drab marbles before the game begins (leaving one red and three drab). Would player I be well advised to accept the deal?* (Answer: *No*; a *yes*-response shows aptitude for Hollywood script writing.)

All three of the above examples can, of course, be easily solved by means of a probability tree (as in FIGURE 1 with $r = 1$). Alternatively, think of the games as consisting of n rounds with the decisive outcome being assigned at random to one of the rounds: thus the first payer in the *Gunsmoke* duel chooses three of the six shot glasses, giving him three chances out of six to drink the strychnine; likewise the first player has three chances out of five to win the red marble in problem A, and 2 of 4 to win it in problem B.

In this paper we explore a natural generalization of these simple problems, a generalization that involves such basic topics as probability trees, problem solving, and recursive formulas. Among the noteworthy features, not only does problem B offer an unhackneyed example of a probability question that for many would clash with their intuition, but the general questions provide a nontrivial application of adding fractions the way students seem to prefer to do it, adding the numerators and adding the denominators.

MAIN PROBLEM. *An urn contains r red and d drab marbles. Two players take turns drawing a marble from the urn; what is the probability that player I (the player drawing first) is the first to draw a red marble?*

One approach to this problem is by way of a probability tree:

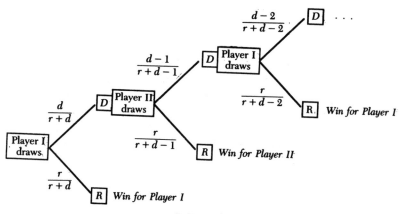

FIGURE 1
The main problem's probability tree.

For $i = 1, 2$, $x = d, d - 1, \ldots, 0$, and $y = r, r - 1, \ldots, 0$, let $P_i(x, y)$ be the probability of a win by the player who draws the ith marble in a game that starts with x drabs and y reds. Then, as indicated in FIGURE 1,

$$
\begin{aligned}
P_1(d, r) &= \frac{r}{r+d} + \frac{d}{r+d} P_2(d-1, r) \\
&= \frac{r}{r+d} + \frac{d}{r+d}(1 - P_1(d-1, r)) \\
&= 1 - \frac{d}{r+d} P_1(d-1, r) \\
&= 1 - \frac{d}{r+d} + \frac{d(d-1)}{(r+d)(r+d-1)} P_2(d-2, r), \text{ etc.} \quad (1)
\end{aligned}
$$

Consequently,

$$
P_1(d, r) = \frac{\binom{r+d}{r} - \binom{r+d-1}{r} + \binom{r+d-2}{r} - \cdots + (-1)^d \binom{r}{r}}{\binom{r+d}{r}}. \quad (2)
$$

Another way to obtain a formula for $P_1(d, r)$ makes use of the observation that the probability of choosing the first red marble on the kth round has a negative *hypergeometric distribution*; that is

$$
\text{Prob } (X = k) = \frac{\binom{d}{k-1}}{\binom{r+d}{k-1}} \cdot \frac{r}{r+d-k+1} = \frac{\binom{r+d-k}{r-1}}{\binom{r+d}{r}},
$$

a formula that can be found in standard references such as [4, p. 157] along with its expected value and variance,

$$
\text{E}(d, r) = \frac{r+d}{r+1}, \text{ and } V(d, r) = \frac{rd(r+d+1)}{(r+1)^2(r+2)}.
$$

Since player I wins if the first red marble appears on an odd-numbered draw,

$P_1(d, r) = \sum_{j < (r+d+1)/2} \text{Prob}(X = 2j - 1)$. Pascal's identity

$$\binom{n-1}{r-1} + \binom{n-1}{r} = \binom{n}{r}$$

can be used to transform this sum into formula (2).

Before looking at yet another approach to the problem, we pause for a brief geography break. Towns across the Canadian Prairies are blessed with a freedom from obstacles such as lakes, rivers, hills, and trees. Consequently, the typical town plan is pretty much as in FIGURE 2—roads run north-south (spaced 16 to the mile) and east-west (spaced 9 to the mile). The invigorating climate devastates the pavement each winter, so that each spring finds crews out repaving the roads. This naturally leads to the next problem.

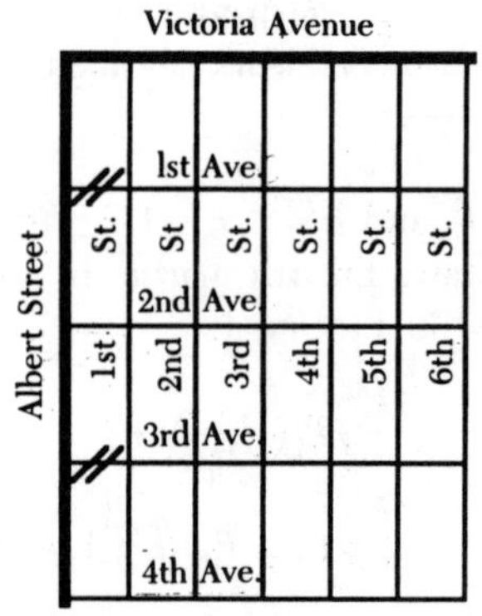

FIGURE 2

Road map for the typical Canadian prairie city; road work is indicated on the first block of all odd numbered avenues.

ALTERNATIVE PROBLEM. *One spring day the crews are at work repairing potholes in all odd-numbered avenues on the first block* (between Albert and 1st Street). *A stranger to town wants to drive from the corner of Victoria and Albert to the corner of dth avenue and rth street* (i.e. making *d* moves down the map, and *r* moves to the right). *What is the probability that, by chance alone, he chooses a route that avoids delays caused by the work crew?*

We learned from George Pólya (see [5, §3.5] for example) how to handle such a problem: If there are *s* ways to get to the corner labelled $(i - 1, j)$ and there are *t* ways to get to $(i, j - 1)$, then there are $s + t$ ways to get to (i, j). Ignoring the road work, there are $\binom{r+d}{r}$ routes to the corner (d, r). These numbers are indicated in bold face in Table 1. The number of ways to get to (i, j) by avoiding the road work can be computed inductively (in italics in the table); there is still only one way to get to $(0, r)$, but the number of ways to get to $(d, 1)$ is $[d/2] + 1$ (where $[d/2]$ is the greatest integer in $d/2$).

The table can be completed using Pólya's rule: When out of the *t* possible routes to $(d - 1, r)$ there are *s* routes that avoid the road work, and out of *v* possible routes on the way to $(d, r - 1)$ *u* of them avoid road work, then the probabilities of avoiding road work are, respectively,

$$P(d - 1, r) = \frac{s}{t} \quad \text{and} \quad P(d, r - 1) = \frac{u}{v};$$

consequently, the probability of arriving at (d, r) without having encountered any road work is

$$P(d, r) = \frac{s + u}{t + v}.$$

This so-called *mediant of* s/t *and* u/v, long used by students to drive up the blood pressure of mathematics teachers around the world, is also a fundamental operation in the study of Farey series [2, Chapter 3].

The alternative problem above is, of course, a disguised version of the main problem. An easy way to see the equivalence is to note that each path to (d, r) corresponds to a sequence of d Ds and r Rs. Of the $\binom{r+d}{r}$ paths to (d, r), those that avoid road work correspond to sequences whose first R appears after an even number (possibly zero) of Ds. Evidently, choosing a D-R sequence at random is equivalent to continuing to play the game described in the main problem (after the winner has been decided) until the urn is empty.

TABLE 1

The number of routes to the corner of dth avenue and rth street is given in bold face, while the number of routes that avoid road work is in italics; their quotient (which equals the probability of avoiding the road work) is also provided.

	$r = 1$		$r = 2$		$r = 3$		$r = 4$		$r = 5$	
$d = 0$	$\frac{1}{1}$	1.000	$\frac{1}{1}$	1.000	$\frac{1}{1}$	1.000	$\frac{1}{1}$	1.000	$\frac{1}{1}$	1.000
$d = 1$	$\frac{1}{2}$	0.500	$\frac{2}{3}$	0.667	$\frac{3}{4}$	0.750	$\frac{4}{5}$	0.800	$\frac{5}{6}$	0.833
$d = 2$	$\frac{2}{3}$	0.667	$\frac{4}{6}$	0.667	$\frac{7}{10}$	0.700	$\frac{11}{15}$	0.733	$\frac{16}{21}$	0.762
$d = 3$	$\frac{2}{4}$	0.500	$\frac{6}{10}$	0.600	$\frac{13}{20}$	0.650	$\frac{24}{35}$	0.686	$\frac{40}{56}$	0.714
$d = 4$	$\frac{3}{5}$	0.600	$\frac{9}{15}$	0.600	$\frac{22}{35}$	0.629	$\frac{46}{70}$	0.657	$\frac{86}{126}$	0.683
$d = 5$	$\frac{3}{6}$	0.500	$\frac{12}{21}$	0.571	$\frac{34}{56}$	0.607	$\frac{80}{126}$	0.635	$\frac{166}{252}$	0.659

We conclude with four further observations concerning these probabilities.

1. $P_1(d, r) \geq P_2(d, r)$; that is, *the first player to draw has at worst an even chance of winning.* To see this, note that any D-R sequence corresponds to a win for player II if, and only if, the first R appears in an even position (after an odd number of Ds). By switching that R with the preceding D one sees that for each such sequence there corresponds a winning sequence for player I. Since this correspondence represents an injection from the set of II's winning sequences into the set of I's winning sequences, player I has at least as many chances to win as his opponent, as claimed. (Indeed, it is easy to check that I has more winning sequences—$P_1(d, r) > P_2(d, r)$—when d is even or $r > 1$.) As an immediate consequence, when "winning" means that you choose the strychnine, we advise that you allow your opponent to go first, despite what they tell you on television.

2. The behavior of the probabilities $P(d, 1)$ that produces the measure of surprise in problem B (namely, *the sequence* $P(d, 1)$, $d = 1, 2, \ldots,$ *is not monotonic*) disap-

pears for $r > 1$. More precisely, *for $r > 1$ $P_1(d, r)$ decreases monotonically with r fixed, and it increases monotonically with d fixed.* This follows immediately from the basic property of mediants (namely $(s/t) > (u/v)$ implies $(s/t) > (s + u)/(t + v) > (u/v)$), after noting that $P_1(0, r) = 1$ for all r while $P_1(d, 2)$ is nonincreasing (since for d even, $P_1(d - 1, 2) = P_1(d, 2) = P_1(d, 1) = [d/2] + 1/d + 1$, a number that decreases with d.

The sequences of the numerators that appear in each column with $r > 2$ evidently lack wide popularity. In fact, only the columns $r = 2$ and $r = 3$ appear in Sloane's list of sequences [6]. Both sequences (1, 2, 4, 6, 9, 12,...) and (1, 3, 7, 13, 22,...) apparently arise from a study of partition functions [1].

3. *For all r*, $\lim_{d \to \infty} P_1(d, r) = 1/2$. This is seen immediately for $r = 1$ by computing $P_1(d, 1) = [d/2] + 1/d + 1$. When $r > 1$ the claim follows from observation 2 above (which states that $P_1(d, r)$ is decreasing) and formula (2) (which says that $P_1(d, r) = r/r + d + d/r + d P_2(d - 1, r)$). As an immediate by-product, interpreting this limit in terms of formula (2) (and remembering that $r > 0$) we conclude that for large d,

$$\sum_{j=0}^{d} (-1)^j \binom{r + j}{j} \approx \frac{(-1)^d}{2} \binom{r + d}{r}.$$

The sum on the left can be recognized as (1.130) in Gould's table of combinatorial identities [3, p. 17].

4. *For all d*, $\lim_{r \to \infty} P_1(d, r) = 1$. In fact, the probability that player I wins on the first draw (namely $r/(d + r)$) approaches 1 as r goes to infinity.

REFERENCES

1. Evelyn Fix and J. L. Hodges, Jr., Significance probabilities of the Wilcoxon test, *Annals of Math. Stats.*, 26 (1955) 301–312.
2. G. H. Hardy and E. M. Wright, *An Introduction to the Theory of Numbers*, 4th edition, Oxford University Press, Fair Lawn, NJ, 1962.
3. Henry W. Gould, *Combinatorial Identities*, Morgantown Printing and Binding Co., Morgantown, WV, 1972.
4. Norman Lloyd Johnson and Samuel Kotz, *Discrete Distributions*, Houghton Mifflin Co., Boston, 1969.
5. George Pólya, *Mathematical Discovery*, John Wiley & Sons, Inc., New York, 1981.
6. Sloane, N. J. A., *A Handbook of Integer Sequences*, Academic Press, New York, 1973.

A New Elementary Proof of Stirling's Formula

C. L. FRENZEN
Naval Postgraduate School
Monterey, CA 93943-5100

If n is a positive integer, the ratio of the $(k+1)$st term to the kth term in the power series

$$e^n = 1 + n + \frac{n^2}{2!} + \cdots \tag{1}$$

is n/k. Thus, the sequence of terms increases as long as $k < n$ and decreases when $k > n$. The nth and $(n+1)$st terms have the same magnitude, $n^n/n!$, and this is the largest magnitude possible for any term in the series in (1). What is the behavior of the ratio of e^n to the largest term in its power series as $n \to \infty$? This question is answered by Stirling's formula, usually written in the form

$$n! \sim n^n e^{-n} \sqrt{2\pi n} \quad (n \to \infty). \tag{2}$$

Equation (2) means that

$$\lim_{n \to \infty} \frac{n!}{n^n e^{-n} \sqrt{2\pi n}} = 1.$$

More generally, f is said to be asymptotic to g as $n \to \infty$, written $f(n) \sim g(n)$ $(n \to \infty)$, if

$$\lim_{n \to \infty} \frac{f(n)}{g(n)} = 1.$$

To answer the question above, we can write (2) as

$$e^n \frac{n!}{n^n} \sim \sqrt{2\pi n} \quad (n \to \infty). \tag{3}$$

Stirling's formula appears in many different disciplines, from algorithm analysis to statistical mechanics. Many derivations of it have been given. See, for example, [1], [2], and [3]. In this note we present a new elementary proof of (2); on the way we will see that the two largest terms in the power series for e^n asymptotically separate its sum into equal parts as $n \to \infty$. We begin with the power series for e^n in (1) and pare it down to a point where we can conclude (3). The proof requires little beyond first-year calculus. More specifically, it requires the following three results:

$$e^x > 1 + x, \quad (x \neq 0). \tag{4}$$

This is easily proved by noting that

$$\int_0^x (1 - e^{-t}) \, dt > 0, \quad x \neq 0.$$

If $x > -1$, $x \neq 0$, and $m > 1$ is an integer, then

$$(1 + x)^m > 1 + mx. \tag{5}$$

This result, often called Bernoulli's inequality, is proved easily by induction. The evaluation of the definite integral

$$\int_0^\infty e^{-x^2}\,dx = \frac{\sqrt{\pi}}{2}, \tag{6}$$

not quite within reach of most first-year calculus students, can be performed in the standard way by considering an iterated double integral and changing to polar coordinates (see, for example, [1, p. 128, Exercise 10]).

For $n \geq 1$, the terms in the power series for e^n in (1) are of two types: those of the form $n^{n+k}/(n+k)!$ with $k \geq 0$ and those of the form $n^{n-k-1}/(n-k-1)!$ with $0 \leq k \leq n-1$. When $k = 0$ each of these expressions has magnitude $n^n/n!$. To compare their magnitudes to $n^n/n!$ when $k > 0$, we let

$$\frac{n^{n-k-1}}{(n-k-1)!} = \frac{n^n}{n!}\alpha_k \qquad \frac{n^{n+k}}{(n+k)!} = \frac{n^n}{n!}\beta_k,$$

where

$$\alpha_k = \left(1 - \frac{0}{n}\right)\left(1 - \frac{1}{n}\right)\cdots\left(1 - \frac{k}{n}\right) \tag{7}$$

for $0 \leq k \leq n-1$, and

$$\begin{aligned}\beta_k &= \frac{1}{\left(1 + \dfrac{0}{n}\right)\left(1 + \dfrac{1}{n}\right)\cdots\left(1 + \dfrac{k}{n}\right)} \\ &= \left(1 - \frac{0}{n+0}\right)\left(1 - \frac{1}{n+1}\right)\cdots\left(1 - \frac{k}{n+k}\right)\end{aligned} \tag{8}$$

for $0 \leq k < \infty$. We can then write

$$e^n\frac{n!}{n^n} = \sum_{k=0}^{n-1}\alpha_k + \sum_{k=0}^{\infty}\beta_k. \tag{9}$$

Note that $\alpha_k \leq \beta_k \leq 1$ for $0 \leq k \leq n-1$, and that α_k and β_k are both monotone decreasing functions of k. Using (4) and (8), we can show that

$$\beta_k \leq e^{-\left(\frac{0}{n+0} + \frac{1}{n+1}\cdots + \frac{k}{n+k}\right)}. \tag{10}$$

Now

$$\frac{0}{n+0} + \frac{1}{n+1}\cdots + \frac{k}{n+k} \geq \frac{0+1+\cdots+k}{n+k} \geq \frac{k^2}{2(n+k)}, \tag{11}$$

so that when $k \geq n$ we have $(1/2)k^2/(n+k) \geq k/4$ and (10) and (11) imply

$$\sum_{k=n}^{\infty}\beta_k \leq \sum_{k=n}^{\infty}e^{-k/4} = e^{-n/4}\sum_{k=0}^{\infty}e^{-k/4}. \tag{12}$$

On the other hand, when $k \leq n$, we have $(1/2)k^2/(n+k) \geq k^2/4n$ and (10) and (11) imply

$$\alpha_k \leq \beta_k \leq e^{-k^2/4n}, \quad (0 \leq k \leq n-1). \tag{13}$$

This suggests the existence of a k_n between 0 and $n-1$ beyond which α_k and β_k

are asymptotically unimportant. To make this more precise, for $0 < \varepsilon < 1/2$ we define $k_n = \lfloor n^{(1/2)+\varepsilon} \rfloor$, the greatest integer less than or equal to $n^{(1/2)+\varepsilon}$. The integer k_n satisfies the inequalities

$$n^{(1/2)+\varepsilon} - 1 < k_n \leq n^{(1/2)+\varepsilon}.$$

We will see how to select ε to our advantage. For $k_n < k < n - 1$, (13) implies that

$$\alpha_k \leq \beta_k < \beta_{k_n+1} < e^{-n^{2\varepsilon}/4}.$$

Summing these inequalities on k then yields

$$\sum_{k=k_n+1}^{n-1} \alpha_k \leq \sum_{k=k_n+1}^{n-1} \beta_k \leq n e^{-n^{2\varepsilon}/4}. \tag{14}$$

A combination of (9), (12), and (14) gives, upon letting $n \to \infty$,

$$e^n \frac{n!}{n^n} \sim \sum_{k=0}^{k_n} \alpha_k + \sum_{k=0}^{k_n} \beta_k \qquad (n \to \infty). \tag{15}$$

The next step is to determine the asymptotic behavior of the right side of (15). From (4) it follows that $1 - x \leq e^{-x} \leq 1/(1+x)$ for $x \geq 0$, and using this in (7) and (8) yields

$$\alpha_k \leq e^{-k(k+1)/2n} \leq \beta_k, \qquad (0 \leq k \leq k_n).$$

This in turn implies

$$\alpha_k \leq e^{-k^2/2n} \leq e^{k_n/2n} \beta_k, \qquad (0 \leq k \leq k_n). \tag{16}$$

Now (7) and (8) also show that

$$\frac{\alpha_k}{\beta_k} = \left[1 - \left(\frac{0}{n}\right)^2\right]\left[1 - \left(\frac{1}{n}\right)^2\right] \cdots \left[1 - \left(\frac{k}{n}\right)^2\right]$$

$$\geq \left[1 - \left(\frac{k}{n}\right)^2\right]^k$$

$$\geq 1 - \frac{k^3}{n^2}, \tag{17}$$

for $0 \leq k \leq k_n$, where (5) has been used to obtain the last inequality in (17). Since $\alpha_k \leq \beta_k$ for $0 \leq k \leq k_n$, (17) implies

$$\left[1 - \frac{k_n^3}{n^2}\right]\beta_k \leq \alpha_k \leq \beta_k \tag{18}$$

when $0 \leq k \leq k_n$. The inequalities in (16) and (18) can then be combined and summed from $k = 0$ to $k = k_n$ to give

$$\left[1 - \frac{k_n^3}{n^2}\right]\sum_{k=0}^{k_n} \beta_k \leq \sum_{k=0}^{k_n} \alpha_k \leq \sum_{k=0}^{k_n} e^{-k^2/2n} \leq e^{k_n/2n}\sum_{k=0}^{k_n} \beta_k. \tag{19}$$

Since $k_n/n \leq n^{(-1/2)+\varepsilon}$ and $k_n^3/n^2 \leq n^{(-1/2)+3\varepsilon}$, if we now choose ε so that $0 < \varepsilon < 1/6$, then $k_n/n \to 0$ and $k_n^3/n^2 \to 0$ as $n \to \infty$. We divide each term in (19) by

$\sum_{k=0}^{k_n} \beta_k$ and let $n \to \infty$ to conclude that

$$\sum_{k=0}^{k_n} \alpha_k \sim \sum_{k=0}^{k_n} e^{-k^2/2n} \quad \text{and} \quad \sum_{k=0}^{k_n} \beta_k \sim \sum_{k=0}^{k_n} e^{-k^2/2n}$$

as $n \to \infty$. Consequently (15) becomes

$$e^n \frac{n!}{n^n} \sim 2 \sum_{k=0}^{k_n} e^{-k^2/2n} \quad (n \to \infty). \tag{20}$$

Together with (14) these equations imply that

$$e^n \frac{n!}{n^n} \sim 2 \sum_{k=0}^{n-1} \alpha_k \quad (n \to \infty). \tag{21}$$

If we employ the definition of α_k given in (7) and rearrange (21), we obtain

$$1 + \frac{n}{1!} + \frac{n^2}{2!} + \cdots + \frac{n^{n-1}}{(n-1)!} \sim \frac{e^n}{2} \quad (n \to \infty).$$

Thus we have the remarkable fact that *the two largest terms in the power series for e^n asymptotically separate its sum into equal parts as $n \to \infty$!*

To estimate the sum on the right side of (20) asymptotically, note that the function f expressed by $f(x) = e^{-x^2}$ is positive and monotonically decreasing on $[0, \infty)$. If we set $h = 1/\sqrt{2n}$, it follows that

$$\int_h^{(k_n+1)h} f(x)\,dx \leq \sum_{k=1}^{k_n} hf(kh) \leq \int_0^{k_n h} f(x)\,dx.$$

Since $k_n h \to \infty$ and $h \to 0$ as $n \to \infty$, we let $n \to \infty$ to obtain

$$\lim_{n \to \infty} \frac{1}{\sqrt{2n}} \sum_{k=1}^{k_n} e^{-k^2/2n} = \int_0^\infty e^{-x^2} dx,$$

or, using (6),

$$\sum_{k=1}^{k_n} e^{-k^2/2n} \sim \frac{\sqrt{2\pi n}}{2} \quad (n \to \infty).$$

It is also clear that

$$\sum_{k=0}^{k_n} e^{-k^2/2n} \sim \frac{\sqrt{2\pi n}}{2} \quad (n \to \infty), \tag{22}$$

so that by combining (20) and (22) we obtain (3) and hence (2).

REFERENCES

1. Tom M. Apostol, *Calculus*, Vol. 2, Blaisedell Pub. Co., Waltham, MA, 1962.
2. A. J. Coleman, A simple proof of Stirling's formula, *Amer. Math. Monthly* 58 (1951) 334–336.
3. Serge Lang, *A First Course in Calculus*, 4th edition, Addison-Wesley Publishing Co., Reading, MA, 1978.

The Product of Chord Lengths of a Circle

ANDRE P. MAZZOLENI
Texas Christian University
Fort Worth, TX 76129

SAMUEL SHAN-PU SHEN
University of Alberta
Edmonton, Canada T6G 2G1

As anyone who has ever had any exposure to complex variables is aware, the subject is full of surprises. From applications in fluid dynamics to the closed-form summation of infinite series, complex analysis has applications that would seem to have no relation to a theory concerned with the "imaginary" number $\sqrt{-1}$. This paper presents an intriguing result in geometry that can be derived by setting the problem in the complex plane and applying the theory of residues (see also [2], p. 69, problem 44).

Suppose we have a circle of unit radius, whose circumference is divided into 8 equal arcs by 8 points. Suppose also that we draw lines from one of the points to each of the other seven points as shown in FIGURE 1. Consider then the problem of determining the product of the 7 chord lengths. It turns out that this product is equal to the number of points we started with, namely 8. In fact, as we shall show below, if we start with n points, the product of the $n-1$ chord lengths we construct will *always* be equal to n. One can easily check this result for the cases $n = 2, 3,$ and 4.

For the general result, suppose we have a circle of unit radius and n points that divide the circumference into n equal arcs. Let $c_1, c_2, \ldots, c_{n-1}$ denote chords drawn from one of the points to each of the remaining $n-1$ points (see FIGURE 1). The product of the $n-1$ chord lengths is just n, i.e.

$$\prod_{k=1}^{n-1} |c_k| = n, \tag{1}$$

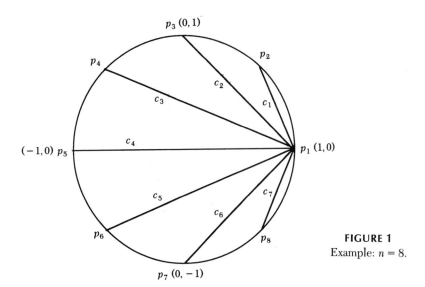

FIGURE 1
Example: $n = 8$.

where $|c_k|$ denotes the length of the chord $c_k, k = 1, \ldots, n-1$ (see also [3], p. 32, problem 160 and [1], pp. 33–34, problems 4.19, 4.20 for related results).

Proof. Without loss of generality, let the originating point be the point $(1, 0)$ in the complex plane. Then the n points can be represented in the complex plane by

$$p_k = e^{i\frac{2\pi(k-1)}{n}}.$$

Since chord c_k is the line from p_1 to p_{k+1}, we have

$$\prod_{k=1}^{n-1} |c_k| = \prod_{k=1}^{n-1} \left| 1 - e^{i\frac{2\pi k}{n}} \right|. \tag{2}$$

Consider the function

$$f(z) = \frac{1}{z^n - 1} = \frac{1}{\displaystyle\prod_{k=1}^{n} \left(z - e^{i\frac{2\pi k}{n}} \right)}.$$

The calculation of the residue of f at $z = 1$ can be done by using the formula

$$\mathrm{Res}(f, 1) = \lim_{z \to 1} (z - 1)f(z) = \frac{1}{\displaystyle\prod_{k=1}^{n-1} \left(1 - e^{i\frac{2\pi k}{n}} \right)}. \tag{3}$$

Since f has a simple pole at $z = 1$, $\mathrm{Res}(f, 1)$ can also be calculated by

$$\mathrm{Res}(f, 1) = \frac{1}{\dfrac{d}{dz}(z^n - 1)\Big|_{z=1}} = \frac{1}{n}. \tag{4}$$

Hence $\prod_{k=1}^{n-1} |c_k| = n$.

REFERENCES

1. R. P. Boas, *Invitation to Complex Analysis*, Random House Inc., New York, 1987.
2. E. Kreyszig, *Solution Manual for Advanced Engineering Mathematics*, 6th edition, John Wiley & Sons, Inc., New York, 1988.
3. M. R. Spiegel, *Shaum's Outline Series: Theory and Problems of Complex Variables*, McGraw-Hill Book Co., New York, 1964.

Four Collinear Griffiths Points

JORDAN TABOV
Bulgarian Academy of Sciences
1113 Sofia, Bulgaria

It is natural to expect that a theorem discovered independently by several mathematicians is in some sense remarkable. Such a theorem is mentioned by R. Johnson on pp. 245 and 246 of his classical book *Advanced Euclidean Geometry* [1]. This theorem, which we shall use here as the basis for a new result, was published by Griffiths in 1857 (references are given in [1]), was rediscovered by Weil in 1880 and by McCay in 1889 (see [2] and [3], respectively). It reads:

When a point P moves along a line through the circumcenter of a given triangle △, the circumcircle of the pedal triangle of P with respect to △ passes through a fixed point on the nine-point circle of △.

The *pedal triangle* of a point P with respect to an arbitrary triangle $A_1A_2A_3$ is shown in FIGURE 1. Its vertices are C_1, C_2 and C_3, the feet of the perpendiculars from P to A_2A_3, A_3A_1, and A_1A_2, respectively. When P coincides with the circumcenter of △ $A_1A_2A_3$, then clearly C_1, C_2, and C_3 are the midpoints of the sides A_2A_3, A_3A_1, and A_1A_2, respectively. The circle passing through these midpoints is called the *nine-point circle* of △ $A_1A_2A_3$. It is well known (see e.g. [1] and [4]) that it also passes through the feet of the altitudes of △ $A_1A_2A_3$ and through the midpoints of the segments joining the orthocenter, H, of △ $A_1A_2A_3$ with vertices A_1, A_2, and A_3 (see FIGURE 2).

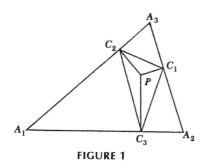

FIGURE 1

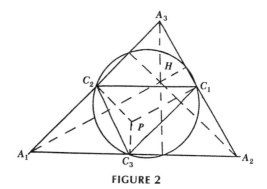

FIGURE 2

This theorem allows us to associate a specially placed point—we call it a *Griffiths point*—with every pair $(\triangle, l)$ of a triangle $\triangle$ and a line l through its circumcenter. Namely, the Griffiths point is the fixed point described in Griffiths' theorem.

In FIGURE 3, a line l passes through the circumcenter O of the triangle $A_1 A_2 A_3$; P is an arbitrary point on l. The point W_4 is the Griffiths point for the pair $(\triangle A_1 A_2 A_3, l)$. It lies on the circumcircles of the pedal triangles of the points O and P (and, according to Griffiths' theorem, of any point on l) with respect to $\triangle A_1 A_2 A_3$.

Four points A_1, A_2, A_3, and A_4 on a circle and a line l through the center of this circle generate four pairs $(\triangle_i, l)$, $i = 1,:2, 3, 4$, where $\triangle_1 \equiv \triangle A_2 A_3 A_4$, $\triangle_2 \equiv A_3 A_4 A_1$, and so on, and thus four corresponding Griffiths points W_i, $i = 1, 2, 3, 4$. Our purpose in this note is to state and prove a new result.

THEOREM. W_1, W_2, W_3, and W_4 *are collinear.*

The proof uses complex numbers. Along the way we shall prove Griffiths' theorem as well.

Let us fix an *Argand diagram*, whose origin coincides with the center O of the unit circle Γ passing through A_1, A_2, A_3, and A_4 such that the real axis lies on l.

By $a_1, w, \ldots$ we shall denote the complex numbers, corresponding respectively to $A_1, W, \ldots$.

Let P be an arbitrary point on l and let C_1, C_2, and C_3 be the feet of the perpendiculars from P to the lines $A_2 A_3$, $A_3 A_1$, and $A_1 A_2$, respectively (see FIGURE 3).

LEMMA.

$$c_1 = \tfrac{1}{2}(a_2 + a_3 + p - p a_2 a_3). \tag{1}$$

Proof. Since $C_1 \in A_2 A_3$, then

$$\frac{c_1 - a_2}{a_3 - a_2} = \frac{\bar{c}_1 - \bar{a}_2}{\bar{a}_3 - \bar{a}_2}.$$

(This equality is equivalent to the fact that the ratio $(c_1 - a_2)/(a_3 - a_2)$ is a real number, i.e. that the argument of the numbers $c_1 - a_2$ and $a_3 - a_2$ differ by either $0°$ or $180°$), which reduces (having in mind that $|a_2| = |a_3| = 1$) to

$$a_2 + a_3 = c_1 + a_2 a_3 \bar{c}_1. \tag{2}$$

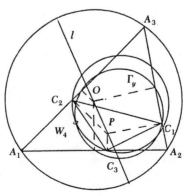

FIGURE 3

And since $PC \perp A_2A_3$, then

$$(p - \bar{c}_1)(a_2 - a_3) = (p - c_1)(\bar{a}_3 - \bar{a}_2).$$

This equality is equivalent to the fact that the ratio $(p - c_1)/(a_2 - a_3)$ is purely imaginary, i.e. that the arguments of the numbers $p - c_1$ and $a_2 - a_3$ differ by either 90° or 270° and reduces to

$$\bar{c}_1 = p - \frac{p - c_1}{a_2 a_3}. \tag{3}$$

Now (2) and (3) yield (1), which completes the proof of the lemma.

Similarly $c_2 = \frac{1}{2}(a_3 + a_1 + p - pa_3 a_1)$ and $c_3 = \frac{1}{2}(a_1 + a_2 + p - pa_1 a_2)$.

Denote by W_4 the point defined by

$$w_4 = \frac{1}{2}(a_1 + a_2 + a_3 - a_1 a_2 a_3). \tag{4}$$

We will show that the points C_1, C_2, C_3, and W_4 are cyclic, i.e. that W_4 lies on the circumcircle of $\triangle C_1 C_2 C_3$. In fact this proves Griffiths' theorem, because W_4 does not depend on the position of P.

In order to do this we must show that

$$\frac{w_4 - c_2}{c_1 - c_2} : \frac{w_4 - c_3}{c_1 - c_3} = \frac{\bar{w}_4 - \bar{c}_2}{\bar{c}_1 - \bar{c}_2} : \frac{\bar{w}_4 - \bar{c}_3}{\bar{c}_1 - \bar{c}_3}. \tag{5}$$

(This equality is equivalent to the fact that the sum of two opposite angles of the quadrilateral with vertices at C_1, C_2, C_3, and W_4 is equal to 180°.)

From the above lemma and (4) we obtain

$$w_4 - c_2 = \frac{1}{2}(a_1 + a_2 + a_3 - a_1 a_2 a_3) - \frac{1}{2}(a_3 + a_1 + p - pa_3 a_1)$$
$$= \frac{1}{2}(p - a_2)(a_3 a_1 - 1),$$

$$\bar{w}_4 - \bar{c}_2 = \frac{1}{2}(p - \bar{a}_2)(\bar{a}_3 \bar{a}_1 - 1)$$
$$= \frac{1}{2}\left(p - \frac{1}{a_2}\right)\left(\frac{1}{a_3}\frac{1}{a_1} - 1\right) = -\frac{1}{2a_1 a_2 a_3}(pa_2 - 1)(a_3 a_1 - 1),$$

$$c_1 - c_2 = \frac{1}{2}(a_1 - a_2)(pa_3 - 1),$$

$$\bar{c}_1 - \bar{c}_2 = \frac{1}{2}(\bar{a}_1 - \bar{a}_2)(p\bar{a}_3 - 1) = -\frac{1}{2a_1 a_2 a_3}(a_1 - a_2)(p - a_3),$$

and similarly for $w_4 - c_3$, $\bar{w}_4 - \bar{c}_3$, $c_1 - c_3$, and $\bar{c}_1 - \bar{c}_3$. When substituted in (5) these expressions reduce it to an identity, and hence C_1, C_2, C_3, and W_4 are cyclic.

Since W_4 does not depend on the position of $P \in l$, W_4 is the Griffiths point of the pair $(\triangle_4, l)$. (Note that, as it was pointed out above, the nine-point circle of any triangle $\triangle$ is the circumcircle of the pedal triangle of the circumcenter of $\triangle$ with respect to $\triangle$. Hence $W_4 \in \Gamma_9$, where Γ_9 is the nine-point circle of $\triangle_4$ (FIGURE 3).

In the same way it may be shown that the point W_1 defined by

$$w_1 = \frac{1}{2}(a_2 + a_3 + a_4 - a_2 a_3 a_4)$$

is the Griffiths point of the pair $(\triangle_1, l)$, and similarly for the points W_2 and W_3.

Now we are ready to prove that W_1, W_2, W_3, and W_4 are collinear. It is sufficient to show that

$$\frac{w_1 - w_2}{\bar{w}_1 - \bar{w}_2} = \frac{w_1 - w_3}{\bar{w}_1 - \bar{w}_3} = \frac{w_1 - w_4}{\bar{w}_1 - \bar{w}_4}. \tag{6}$$

Denoting $a_1 a_2 a_3 a_4$ by a we have

$$w_1 - w2 = \tfrac{1}{2}(a_2 + a_3 + a_4 - a_2 a_3 a_4) - \tfrac{1}{2}(a_3 + a_4 + a_1 - a_3 a_4 a_1)$$
$$= \tfrac{1}{2}(a_1 - a_2)(a_3 a_4 - 1),$$
$$\overline{w}_1 - \overline{w}_2 = \tfrac{1}{2}(\overline{a}_1 - \overline{a}_2)(\overline{a}_3 \overline{a}_4 - 1)$$
$$= \frac{1}{2 a_1 a_2 a_3 a_4}(a_1 - a_2)(a_3 a_4 - 1) = \frac{\overline{a}}{2}(w_1 - w_2).$$

Similarly $\overline{w}_1 - \overline{w}_3 = \tfrac{\overline{a}}{2}(w_1 - w_3)$ and $\overline{w}_1 - \overline{w}_4 = \tfrac{\overline{a}}{2}(w_1 - w_4)$; these results prove (6). Hence W_1, W_2, W_3, and W_4 are collinear.

REFERENCES

1. R. A. Johnson, *Advanced Euclidean Geometry*, Dover Publications, Mineola, NY, 1960.
2. M. Weil, Note sur le triangle inscrit et circonscrit a deux coniques, *Nouvelles Annales de Mathématiques*, series 2, 19(1880) 253–261.
3. W. S. McCay, On three similar figures, with an extension of Feuerbach's Theorem, *Trans. Royal Irish Acad.* 29(1889) 303–320.
4. R. Honsberger, *Mathematical Gems* II, MAA, Washington, DC, 1976.

Did you know that ... ?

The journal now called *Mathematics Magazine* was started in 1926 as a series of pamphlets to encourage membership in the Louisiana-Mississippi Section of the MAA. The so-called "Object of the Campaign" was to bring together high school and college teachers of mathematics through closer professional contact via membership in the MAA. The pamphlet listed as endorsers of the campaign the school superintendents of the two states, the presidents of eight institutions of higher learning and the national officers of the MAA.

The Section Chairman, S.T. Sanders, was one of the individuals principally responsible for founding (and funding) the journal. Eight pamphlets appeared in the first year, with the last five of these actually appearing in 1927. After the 1927 Section meeting in Shreveport, at which a Branch of NCTM was founded, Professor Sanders, H.E. Slaught, Past President of the MAA, and P.K. Smith decided to continue the publication of the pamphlets as *The Mathematics News Letter*. The regular price was 50¢ per year for ten issues. In 1934-35, with Volume 9, the name of the journal was changed to *National Mathematics Magazine*.

To find out more...see Edwin Beckenbach's brief history of the journal in *Mathematics Magazine 50 Year Index (Vols. 1-50, 1926-1977)*.

Exploring Complex-Base Logarithms

STEPHEN P. HUESTIS
University of New Mexico
Albuquerque, NM 87131

Students of complex analysis soon discover that the natural logarithm is a multivalued function with an infinite number of branches, reflecting the multiple representation of any complex number. If $z = x + iy \equiv (x, y)$, then in polar form:

$$z = re^{i[\theta + 2\pi n]}$$

with $r = +\sqrt{x^2 + y^2}$, $0 \leqslant \theta \leqslant \tan^{-1}(y/x) < 2\pi$, and n any integer. Hence

$$w = \ln(z) = \ln(r) + i(\theta + 2\pi n).$$

Interesting patterns arise when multivalued logarithms are generalized to arbitrary complex bases. To complex base z_1 the logarithm,

$$w = \log_{z_1}(z_2), \tag{1}$$

is defined by

$$z_2 = z_1^w. \tag{2}$$

Consider now that both z_1 and z_2 have multiple representations:

$$z_1 = \rho e^{i[\phi + 2\pi m]}$$

$$z_2 = re^{i[\theta + 2\pi n]}.$$

Taking the natural logarithm of (2), (1) is equivalent to:

$$w = \frac{\ln(r) + i(\theta + 2\pi n)}{\ln(\rho) + i(\phi + 2\pi m)}. \tag{3}$$

As m and n independently run through all integers, (3) defines an infinite number of points in the complex plane that serve as representations of the logarithm of z_2 to base z_1. What pattern do they describe?

Fix an integer m, and let $C_m = \phi + 2\pi m$. Then, as n varies, (3) gives for $w = x + iy$:

$$x = \left[\ln(r) \cdot \ln(\rho) + C_m \cdot (\theta + 2\pi n)\right] / \left[\ln^2(\rho) + C_m^2\right]$$
$$y = \left[\ln(\rho) \cdot (\theta + 2\pi n) - \ln(r) \cdot C_m\right] / \left[\ln^2(\rho) + C_m^2\right]. \tag{4}$$

Replace $(\theta + 2\pi n)$ by a continuous variable s that, when eliminated from (4), gives

$$y = x \cdot \left[\ln(\rho)/C_m\right] - \ln(r)/C_m. \tag{5}$$

That is, for fixed m, points (in the Cartesian plane) corresponding to the base z_1 logarithm of z_2 fall on a straight line of slope $[\ln(\rho)/C_m]$ and intercept $[-\ln(r)/C_m]$; they are points for which $(s - \theta)/2\pi$ is an integer. Note that all lines defined by (5) pass through the point $A = (\ln(r)/\ln(\rho), 0)$, independently of m.

Now fix n in (3) and let $D_n = (\theta + 2\pi n)$; replace $(\phi + 2\pi m)$ by a continuous variable s to get a parametric expression for the locus of points on which logarithm representations fall as m varies:

$$x = \left[\ln(r) \cdot \ln(\rho) + D_n \cdot s\right] / \left[\ln^2(\rho) + s^2\right]$$
$$y = \left[\ln(\rho) \cdot D_n - \ln(r) \cdot s\right] / \left[\ln^2(\rho) + s^2\right]. \tag{6}$$

Eliminating s from (6) is algebraically more unwieldy than for the fixed m case. Anyone who plots several examples, however, will no doubt speculate that (6) is a circle in the Cartesian plane. Substitute (6) into the general equation for a circle with center (u, v) and radius R:

$$(x - u)^2 + (y - v)^2 = R^2,$$

and solve for u, v, and R by evaluating at three convenient values of s (for example, $s = 0$, $\ln(\rho)D_n/\ln(r)$, and ∞), to show that (6) is equivalent to

$$\left[x - \frac{\ln(r)}{2\ln(\rho)}\right]^2 + \left[y - \frac{D_n}{2\ln(\rho)}\right]^2 = \frac{\ln^2(r) + D_n^2}{4\ln^2(\rho)}. \tag{7}$$

As n varies through integer values, (7) defines a family of circles, each with center on the vertical line $x = \ln(r)/[2\ln(\rho)]$, and passing through the point A. The fact that each circle (7) intersects each line (5) at A, guarantees another intersection point, which is the logarithm representation for the corresponding (m, n) pair.

FIGURE 1 shows the geometry of intersecting lines and circles for $z_1 = (3, 2)$, $z_2 = (2, -1)$, and $|m|, |n| \leq 2$; curves are labeled with their m, n values. The point of common intersection is $A = (.627, 0)$; all other intersection points of individual lines and circles are representations of w. FIGURE 2 is the same for $z_1 = (-1, 3)$, $z_2 = (.1, .1)$, with $A = (-1.699, 0)$.

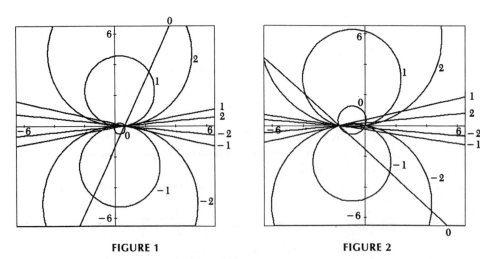

FIGURE 1   FIGURE 2

For fixed m, (4) shows that both $|x|$ and $|y| \to \infty$ as $|n| \to \infty$, so there are logarithm points arbitrarily far from the origin, along each line (5). For fixed n, the circle (7) intersects every line (5); as $|m| \to \infty$ their slopes approach zero, so that the intersection points approach the x (real) axis. FIGURES 3 through 6 display this concentration of logarithm points about the x-axis, for the case $z_1 = (3, 2)$, $z_2 = (2, 1)$. For FIGURE 3, the plot limits are $0 \leq x \leq 2$; $-.1 \leq y \leq .1$ (note distortion), and $|m|, |n| \leq 25$. Each successive figure is a blowup of the small boxed region of the preceding figure. Truncation values for FIGURES 4 through 6 are $|m|, |n| \leq 50$, 200, and 1000, respectively. Truncation at finite m, n leads to the point-free central band of each figure.

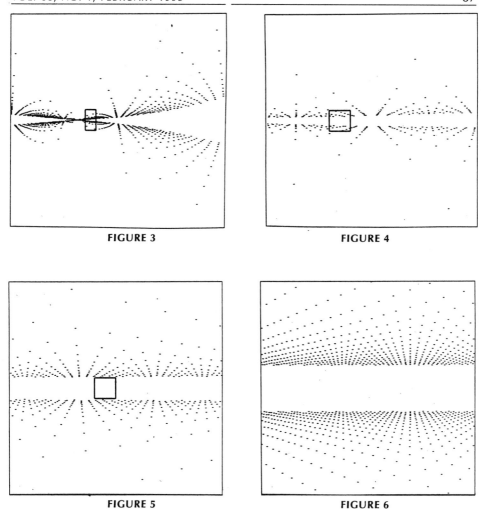

FIGURE 3   FIGURE 4

FIGURE 5   FIGURE 6

This elegant pattern of fractal-like ray structures, upon which complex-base loga-rithms fall, is a particular example of patterns assumed by multiple representations of arbitrary functions of k complex numbers:

$$w = f(z_1, z_2, \ldots, z_k).$$

For example, Gleason [1; p. 324] explicitly comments on the infinite number of representations of

$$w = \ln(z_1 z_2).$$

In this case, however, the pattern is far less interesting. Expressing z_1, z_2 as above,

$$w = \ln(r\rho) + i[\theta + \phi + 2\pi(n + m)]$$

and all representations fall at equal spacing on the single line $x = \ln(r\rho)$.

REFERENCE

1. Andrew M. Gleason, *Fundamentals of Abstract Analysis*, Jones and Bartlett Publishers, Boston, 1991.

PROBLEMS

LOREN C. LARSON, *editor*
St. Olaf College

GEORGE GILBERT, *associate editor*
Texas Christian University

Proposals

To be considered for publication, solutions should be received by July 1, 1995.

1464. *Proposed by Bill Correll, Jr., student, Denison University, Granville, Ohio.*

Find all positive rational numbers $r \neq 1$ such that $r^{1/(r-1)}$ is rational.

1465. *Proposed by Steven W. Knox, University of Illinois, Urbana, Illinois.*

It is a well-known theorem that given any coloring of the plane by two colors, there exists an equilateral triangle with monochromatic vertices. As a generalization, show that given any two-coloring of the plane and any triangle T, there exists a triangle similar to T with monochromatic vertices.

1466. *Proposed by David M. Bloom, Brooklyn College of CUNY, Brooklyn, New York.*

Let m and n be positive integers. If $x_1, \ldots, x_m$ are positive integers whose average is less than $n + 1$ and if $y_1, \ldots, y_n$ are positive integers whose average is less than $m + 1$, prove that some sum of one or more x's equals some sum of one or more y's. (This is a strengthening of Putnam problem A-4, 1993; see this MAGAZINE, April 1994, 156–157.)

1467. *Proposed by John A. Baker, University of Waterloo, Waterloo, Ontario, Canada*

Suppose that $v_0, v_1, \ldots, v_n$ are the vertices of a regular simplex, S, in $\mathbf{R}^n$ centered at the origin. Let

$$v_i = (v_{i1}, v_{i2}, \ldots, v_{in}) \quad \text{for } 0 \leq i \leq n.$$

ASSISTANT EDITORS: CLIFTON CORZAT, BRUCE HANSON, RICHARD KLEBER, KAY SMITH, and THEODORE VESSEY, St. Olaf College and MARK KRUSEMEYER, Carleton College. *We invite readers to submit problems believed to be new and appealing to students and teachers of advanced undergraduate mathematics. Proposals should be accompanied by solutions, if at all possible, and by any other information that will assist the editors and referees. A problem submitted as a Quickie should have an unexpected, succinct solution. An asterisk (∗) next to a problem number indicates that neither the proposer nor the editors supplied a solution.*

Solutions should be written in a style appropriate for Mathematics Magazine. Each solution should begin on a separate sheet containing the solver's name and full address.

Solutions and new proposals should be mailed in duplicate to Loren Larson, Department of Mathematics, St. Olaf College, 1520 St. Olaf Ave., Northfield, MN 55057-1098 or mailed electronically via fax: (507) 663-3549 or e-mail: larson@stolaf.edu.

Prove that, for some $c > 0$,

$$\sum_{i=0}^{n} v_{ij}^2 = c \quad \text{for all } j = 1, 2, \ldots, n.$$

1468. *Proposed by G. Trenkler, University of Dortmund, Dortmund, Germany.*

Let A be a square matrix with real entries satisfying $A^2 = A^T$.
(i) Find its Moore-Penrose inverse A^+ in terms of A.
(ii) Assume A is a 2×2 matrix. Find all solutions to $A^2 = A^T$ that are not symmetric.

Quickies

Answers to the Quickies are on page 74.

Q829. *Proposed by Ismor Fischer, University of Wisconsin, Oshkosh, Wisconsin.*

Show that $1 < (\dfrac{2^x - 1}{x})^{\frac{1}{x-1}} < 2$, for all $x \neq 0, 1$.

Q830. *Proposed by Murray S. Klamkin, University of Alberta, Edmonton, Alberta, Canada.*

Determine $\int_0^1 (1 - x^m)^n \, dx \div \int_0^1 (1 - x^m)^{n-1} \, dx, m, n > 0$, without using beta function integrals.

Q831. *Proposed by Joe J. Rushanan and Pankaj Topiwala, The MITRE Corporation, Bedford, Massachusetts.*

Let $\Sigma_{n,j}$ denote the sum of the kth powers of sums of j distinct elements of $x_1, \ldots, x_n$ of a commutative ring, taken over all possible j-element subsets. For instance, $\Sigma_{3,2} = (x_1 + x_2)^k + (x_1 + x_3)^k + (x_2 + x_3)^k$. Prove that

$$\Sigma_{k,k} - \Sigma_{k,k-1} + \Sigma_{k,k-2} - \cdots + (-1)^{k-1} \Sigma_{k,1} = k! x_1 \cdots x_k.$$

Solutions

Primes, squares, and divisibility   **February 1994**

1438. *Proposed by David M. Bloom, Brooklyn College of CUNY, Brooklyn, New York.*

Let p be a prime with $p > 5$, and let $S = \{p - n^2 : n \in \mathbf{Z}^+, n^2 < p\}$. (For example, if $p = 31$, then $S = \{6, 15, 22, 27, 30\}$.) Prove that S contains two elements a, b such that $1 < a < b$ and a divides b.

Solution by J. S. Frame, Michigan State University, East Lansing, Michigan.

If $p = m^2 + 1 > 5$, then $m > 2$ is even, so $a = p - (m - 1)^2 = 2m$ divides $b = p - 1^2 = m^2 > a$. Otherwise assume $5 < m^2 + 1 < p < (m + 1)^2 - 1$ for some m. Then $1 < a = p - m^2 < 2m$, and for some positive $k < m$, a must divide $(m - k)(m + k) + a = p - k^2 = b$, since each integer between 1 and $2m$ except m is either an $m - k$ or an $m + k$, and $p = m^2 + m$ is impossible.

Also solved by Jack C. Abad, Paul R. Abad, D. F. Bailey, S. F. Barger, Kenneth L. Bernstein, Walter Blumberg, Duane M. Broline, Darin Brown, Glen Van Brummelen (Canada), Con Amore Problem Group (Denmark), Bill Correll, Jr. (student), Robert Gardner, Richard Heeg, Richard Holzsager, Thomas Jager, D. Kipp Johnson, Hans Georg Killingbergtrø (Norway), Libby Krussel and Stephen Penrice, Kee-Wai Lau (Hong Kong), Kari Lehtinen (Finland), Miguel Lerma, Marijo LeVan, Kenneth Levassuer, Paul Li (student), Helen M. Marston, Hugh McGuire (student), Michael Nathanson (student), Taihei Okada, Waldemar Pompe (Poland), Michael Reid (student), F. C. Rembis, Nicholas C. Singer, John S. Sumner, Trinity University Problem Group, Michael Vowe (Switzerland), Sonny Vu (student), A. N.'t Woord (Netherlands), David Zhu, and the proposer.

Holzsager and Killingbergtrø generalized the result to all integers that are not of the form m^2, $m + 1$ and even, $m^2 + m$, and $m^2 + 2m$.

Limit of a spiraling polygon   February 1994

1439. *Proposed by Charles Vanden Eynden, Illinois State University, Normal, Illinois.*

All the lines in the sketch (FIGURE 1) have slope 0, 1, or -1, or are vertical. What point do the points P_n approach?

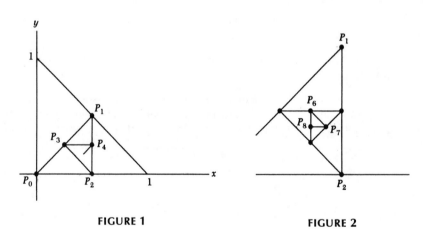

FIGURE 1   FIGURE 2

I. *Solution by Paul Li, student, Harvard University, Cambridge, Massachusetts.*

Set $A = (0, 1)$ and $B = (1, 0)$, and add a few more points to the sequence, as indicated on the diagram FIGURE 2. The coordinates of the first few points are $P_0 = (0, 0)$, $P_1 = (1/2. 1/2)$, $P_2(1/2, 0)$, $P_6 = (3/8, 1/4)$, $P_7 = (7/16, 3/16)$, $P_8 = (3/8, 3/16)$.

At this point we notice that the part of the figure inside the triangle $\triangle P_6 P_7 P_8$ is a scaled-down version of the original figure. If Q is the limit point of P_n, then the triangle $\triangle ABP_0$ is the image of $\triangle P_6 P_7 P_8$ under a dilation about Q. In particular, Q is the intersection point of the lines $P_0 P_8$ and AP_6. From this we find that $Q = (2/5, 1/5)$.

II. *Solution by J. C. Binz, University of Bern, Bern, Switzerland.*

The sequence of points P_n can be described in the complex plane by $p_0 = 0$, $p_1 = (1 + i)/2$, $p_2 = 1/2$, and $p_n = (p_{n-2} + p_{n-3})/2$ for $n \geq 3$.

The corresponding characteristic equation $x^3 - x/2 - 1/2$ has roots $1, d = (-1 + i)/2$, and $\bar{d}$. Hence $p_n = a + bd^n + c\bar{d}^n$, where a, b, c can be determined from p_0, p_1, and p_2. In particular, $a = (2 + i)/5$. Since $|d| < 1$ implies $\lim_{n \to \infty} d^n = 0$, we have $\lim_{n \to \infty} p_n = a$.

Also solved by Aardvark Problem Solving Group, Ralph Abell, Ed Adams, Robert A. Agnew, P. J. Anderson, Michael H. Andreoli, Harry L. Baldwin, Jr., S. F. Barger, Henry J. Barten, Kenneth L. Bernstein, Nirdosh Bhatnagar, J. C. Binz (Switzerland, second solution and generalization), Walter Blumberg, Ada Booth, Mario R. Bordogna, Mark Bowron, Paul Bracken (Canada), Duane M. Broline, David C. Brooks, Darin Brown, Stan Byrd, Cea B. Cristian (Chili), Con Amore Problem Group (Denmark), Bill Correll, Jr. (student), Thomas P. Dence, Charles R. Diminnie, David Doster, Robert L. Doucette, Frank Eccles, Milton P. Eisner, Thomas Vanden Eynden, J. Chris Fisher (Canada), Jesse C. Frey, Sam Gardner (student), Robert Geretschläger (Austria), John F. Goehl, Jr., Jerrold W. Grossman, Arthur Guetter, Dale K. Hathaway, Francis M. Henderson, Richard Holzsager, R. Daniel Hurwitz, Douglas Iannucci, D. Kipp Johnson, David L. Judson, Hans Kappus (Switzerland), Frederick James Kennedy (Canada), Hans Georg Killingbergtrd (Norway), Peter Knapp (student), James M. Kretchmar (student), H. K. Krishnapriyan, Kari Lehtinen (Finland), Nick Lord (England), Robert Mandl, Helen M. Marston, Reiner Martin (student), Allen J. Mauney, Vince McGarry, Gary Miller (Canada), Can Anh Minh and Paul Herman, Stephen Noltie, James D. Nulton, Taihei Okada, Thomas O'Neil, Cornel G. Ormsby, Jeremy Ottenstein (Israel), Ronnie Pavlov and James A. Fitzsimmons, Richard E. Pfiefer, Rob Pratt, F. C. Rembis, Jérémie Rostand (Canada), Henry Schultz, Heinzjürgen Seiffert (Germany), Sherrill Shaffer, Mohammad P. Shaikh, Robert W. Sheets, Nicholas C. Singer, Paul G. Staneski, John S. Sumner, Michael Tehranchi (student), Nora Thornber, Pat Touhey, Trinity University Problem Group, Michael Vowe (Switzerland), Sonny Vu (student), Robert J. Wagner, Doug Wilcock, James W. Wilson, Michael Woltermann, Sam Wood, A. N.'t Woord (Netherlands), Rex H. Wu (student), Robert L. Young, Ted Zerger, Yonggan Zhao, Harald Ziehms (Germany), and the proposer.

Fisher pointed out that a more general version of the problem is discussed in *Crux Mathematicorum*, # 1679, 18:8 (October 1992), 248–250.

r-fold free sets of positive integers

February 1994

1440. *Proposed by Bruce Reznick, Univeristy of Illinois, Urbana, Illinois.*

Let $r \geq 2$ be an integer. We say that a set of positive integers A is *r-fold-free* if $k \in A$ implies $rk \notin A$. Let $f_r(n)$ denote the cardinality of the largest r-fold-free subset of $\{1, 2, \ldots, n\}$. It is easy to calculate $f_r(n)$ for small values of r and n and corresponding sets of maximal cardinality. For example, $f_2(1) = 1, (\{1\})$; $f_2(2) = 1, (\{1\}$ or $\{2\})$; $f_2(3) = 2, (\{1, 3\}$ or $\{2, 3\})$; $f_2(4) = 3, (\{1, 3, 4\})$; $f_2(5) = 4, (\{1, 3, 4, 5\})$.

Prove the following closed formula for $f_r(n)$. Let $[a_m a_{m-1} \cdots a_0]_r$ denote the base r representation of n. Then

$$f_r(n) = \left(\frac{1}{r+1}\right)\left(r \cdot n + \sum_{k=0}^{m} (-1)^k a_k\right).$$

For example, $3000 = [101110111000]_2$, so

$$f_2(3000) = (6000 - 1 + 1 - 1 - 1 + 1 - 1 - 1)/3 = 1999.$$

Solution by Richard Holzsager, The American University, Washington, D.C.

For any $m \leq n$ that is not divisible by r, we have to choose among $m, mr^2, \ldots, mr^k$ (where $mr^k \leq n < mr^{k+1}$) a subset that does not include any two in succession. Clearly one way to maximize the number chosen is to choose alternate ones starting with m. This amounts to throwing away multiples of r, adding back multiples of r^2, throwing out multiples of r^3, and so forth. The overall result is

$$
\begin{aligned}
n &- \lfloor n/r \rfloor + \lfloor n/r^2 \rfloor - \lfloor n/r^3 \rfloor \pm \cdots \\
&= [a_m \cdots a_0]_r - [a_m \cdots a_1]_r + [a_m \cdots a_2]_r - [a_m \cdots a_3]_r \pm \cdots \\
&= a_m \big(r^m - r^{m-1} + \cdots + (-1)^m \big) + \\
&\quad a_{m-1}\big(r^{m-1} - r^{m-2} + \cdots + (-1)^{m-1} \big) + \cdots + a_0 \\
&= \sum_{k=0}^{m} a_k \left(\frac{r^{k+1} + (-1)^k}{r+1} \right) \\
&= \left(\frac{1}{r+1} \right) \left(\sum_{k=0}^{m} a_k r^{k+1} + \sum_{k=0}^{m} (-1)^k a_k \right) = \left(\frac{1}{r+1} \right) \left(r \cdot n + \sum_{k=0}^{m} (-1)^k a_k \right).
\end{aligned}
$$

Also solved by Kenneth L. Bernstein, J. C. Binz (Switzerland), Duane M. Broline, Con Amore Problem Group (Denmark), Bill Correll, Jr. (student), Robert L. Doucette, Thomas Jager, Paul Lambert (student), Paul Li (student), Taihei Okada, A. N.'t Woord (Netherlands), and the proposer.

Convergence of recursive sequence   February 1994

1441. *Proposed by S. C. Woon, Imperial College, London, United Kingdom.*

Let $\pi(0) = 1$ and for each integer $n \geq 1$,

$$
\pi(n) = \sqrt{ 1 + \left(\sum_{k=0}^{n-1} \pi(k) \right)^2 }.
$$

Show that $\displaystyle \lim_{n \to \infty} \frac{2^{n+1}}{\pi(n)} = \pi$.

I. *Solution by Kee-Wai Lau, Hong Kong.*

We first prove by induction that

$$
\pi(n) = \csc\left(\frac{\pi}{2^{n+1}} \right), \qquad \text{for } n = 0, 1, 2, \ldots. \tag{1}
$$

Since $\pi(0) = 1$, the identity is true for $n = 0$. Suppose (1) holds for $n \leq t$. Then by using the recurrence relation for $\pi(n)$, we have

$$
\pi(t+1) = \sqrt{ 1 + \left(\sum_{k=0}^{t} \csc\left(\frac{\pi}{2^{k+1}} \right) \right)^2 }. \tag{2}
$$

Now

$$
\csc\left(\frac{\pi}{2^{k+1}} \right) = \cot\left(\frac{\pi}{2^{k+2}} \right) - \cot\left(\frac{\pi}{2^{k+1}} \right),
$$

and therefore

$$
\sum_{k=0}^{t} \csc\left(\frac{\pi}{2^{k+1}} \right) = \cot\left(\frac{\pi}{2^{t+2}} \right).
$$

Hence by (2), we have $\pi(t+1) = \csc(\pi/2^{t+2})$. This shows that (1) holds for $n = t+1$, and hence, by induction, for all nonnegative integers n.

Since $\lim_{\theta \to 0} \dfrac{\sin \theta}{\theta} = 1$, it easily follows from (1) that

$$\lim_{n \to \infty} \frac{2^n + 1}{\pi(n)} = \pi,$$

as required.

II. *Solution by F. C. Rembis, Clifton, New Jersey.*

From the figure we see that $\theta_1 = \pi/4$ and $\theta_{i+1} = \theta_i/2$. It follows that $\theta_n = \pi/2^{n+1}$ and $\sin \theta_n = 1/\pi(n)$. Thus,

$$\lim_{n \to \infty} 2^{n+1}/\pi(n) = \lim_{n \to \infty} 2^{n+1} \sin\left(\pi/2^{n+1}\right) = \pi.$$

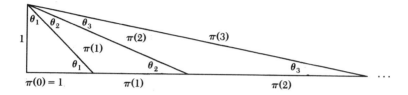

Also solved by *P. J. Anderson (Canada), John Andraos (Canada), Michel Bataille (France), Kenneth L. Bernstein, Nirdosh Bhatnagar, Walter Blumberg, Duane M. Broline, Cea B. Christian (Chili), Con Amore Problem Group (Denmark; two solutions), Bill Correll, Jr. (student), Paul Deiermann and Brian Brill (student), David Doster, Robert L. Doucette, William Heidorn (student), Richard Holzsager, Thomas Jager, Hans Kappus (Switzerland), Hans Georg Killingbergtrø (Norway), Gérard Letac (France), Eugene Levine, Paul Li (student), Nick Lord (England), Reiner Martin (student), Can Anh Minh, Tai-hei Okada, Jeremy Ottenstein (Israel), Michael Reid (student), Harald Schmidt (student, Germany), Heinz-Jürgen Seiffert (Germany), Nicholas C. Singer, John S. Sumner, Michael Tehranchi (student), Michael Vowe (Switzerland), Robert J. Wagner, A. N't Woord (Netherlands), Feng-Shuo Yu, David Zhu,* and the proposer.

Ellipses with one focus   February 1994

1442. *Proposed by W. O. Egerland and C. E. Hansen, Aberdeen Proving Ground, Maryland.*

Prove that two ellipses with exactly one focus in common intersect in at most two points.

Solution by Nick Lord, Tonbridge School, Kent, England.

In polar coordinates, with the common focus as origin, the equations of the two ellipses may be taken as

$$r = \frac{l}{1 + e \cos \theta} \quad \text{and} \quad r = \frac{L}{1 + E \cos(\theta + \alpha)},$$

where not all of $l = L$, $e = E$, $\alpha = 0$ hold.

These meet where

$$\frac{l}{1 + e \cos \theta} = \frac{L}{1 + E \cos(\theta + \alpha)},$$

or,

$$(El \cos \alpha - eL)\cos \theta - (El \sin \alpha)\sin \theta = L - l.$$

Putting the left-hand side into the form "$R \cos(\theta + \beta)$" by setting

$$R = \sqrt{(El \cos \alpha - eL)^2 + (El \sin \alpha)^2} \quad \text{and} \quad \tan \beta = \frac{El \sin \alpha}{El \cos \alpha - eL},$$

the required conclusion follows from the fact that $\cos(\theta + \beta) = (L - l)/R$ has at most 2 solutions for $0 \le \theta < 2\pi$.

Also solved by Michel Bataille (France), Kenneth L. Bernstein, Duane M. Broline, Con Amore Problem Group (Denmark), Wayne Dennis, Ragnar Dybvik (Norway), Michael Golomb, H. Guggenheimer, Joe Howard, Thomas Jager, D. Kipp Johnson, Hans Georg Killingbergtrø (Norway), Helen M. Marston, Stephen Noltie, James D. Nulton and Harry L. Baldwin, Jr., Richard E. Pfiefer, F. C. Rembis, Nora Thornber, Patricia Whalen and Rick Swanson, and the proposers.

Joe Howard and Murray Klamkin pointed out that the problem appeared in Mathematical Gazette, 69.D, (69), 1985, p. 302.

Answers

Solutions to the Quickies on page 69.

A829. For fixed $x \ne 0, 1$, apply the Mean Value Theorem to $f(t) = t^x$ on $[1, 2]$. *Remark.* The lower and upper bounds on the function in the problem are sharp, and represent the limiting values as $x \to \mp \infty$, respectively. Moreover, the function can be extended continuously through the points 0 and 1 by defining it to take on the values $1/\ln 2$ and $4e$, respectively.

A830. Integrating by parts,

$$F_n \equiv \int_0^1 (1 - x^m)^n \, dx = mn \int_0^1 x^m (1 - x^m)^{n-1} \, dx$$

$$= mn \int_0^1 \left((x^m - 1)(1 - x^m)^{n-1} + (1 - x^m)^{n-1} \right) dx$$

or, $F_n(1 + mn) = mn F_{n-1}$. Hence

$$\frac{F_n}{F_{n-1}} = \frac{mn}{1 + mn}.$$

A831. First, consider the identity over the real numbers. Set $x_k = 0$. The left-hand side becomes

$$\Sigma_{k-1, k-1} - (\Sigma_{k-1, k-1} + \Sigma_{k-1, k-2}) + (\Sigma_{k-1, k-2} + \Sigma_{k-1, k-3})$$
$$- \cdots + (-1)^{k-1} \Sigma_{k-1, 1} = 0.$$

Thus, by symmetry, the original sum, which is a homogeneous polynomial of degree k, equals $cx_1 \cdots x_k$. This term arises only from $(x_1 + \cdots + x_k)^k$, and the identity follows. The identity must therefore follow over any commutative ring.

REVIEWS

PAUL J. CAMPBELL, *editor*
Beloit College

Assistant Editor: Eric S. Rosenthal, West Orange, NJ. Articles and books are selected for this section to call attention to interesting mathematical exposition that occurs outside the mainstream of the mathematics literature. Readers are invited to suggest items for review to the editors.

Kolata, Gina, . . . While a mathematician calls classic riddle solved, *New York Times* (National Edition) (27 October 1994) A1, B12. Cipra, Barry, Is the fix in on Fermat's Last Theorem?, *Science* 266 (4 November 1994) 725. Peterson, I., Fermat's famous theorem: Proved at last?, *Science News* (5 November 1994) 295.

Andrew Wiles (Princeton University) has announced a revised proof, with the help of former student Richard Lawrence Taylor (Cambridge University), that overcomes the gap in the proof that he presented in June 1993.

Passell, Peter, Game theory captures a Nobel, *New York Times* (12 October 1994) D1, D6; (National Edition) C1, C6. Nasar, Sylvia, The lost years of a Nobel laureate, *New York Times* (National Edition) (13 November 1994) F1, F8.

The 1994 Nobel Prize in Economic Science ($930,000) was awarded jointly to three pioneers in multi-player game theory: John F. Nash (Princeton University), John C. Harsanyi (UC—Berkeley), and Reinhard Selten (University of Bonn). The award to Nash was particularly poignant; honored for his 27-page Ph.D. thesis on non-cooperative games, he had been ill and unable to work for many years.

Markoff, John, Flaw undermines accuracy of Pentium chips, *New York Times* (24 November 1994) D1, D9. Chip error continuing to dog officials at Intel. (6 December 1994) D9.

Do you have a 90 MHz screamer Pentium PC? Too bad. An error in the numeric coprocessor on the Intel Pentium chip gives wrong answers to some computations. Intel claims that the chip (5 million installed) errs only in rare computations, hence the company offers free replacements of the $500 (originally $1,000) chip only to users whose work demands intensive and correct computation. The error was detected by Intel in June but made public only in November, not by Intel but by mathematician Thomas Nicely (Lynchburg College), who discovered it while computing reciprocals of primes. Intel's 386 and 486 chips also featured errors in arithmetic, which were corrected in subsequent production. As Assistant Editor Eric Rosenthal points out, this story has already "far outgrown" this Reviews column. For further information, there is a collection of "primary source" material at the Worldwide Web site www.mathworks.com, under the "What's New" button, or at the anonymous ftp site ftp.mathworks.com, in directory /pub/tech-support/moler/Pentium.

Pritchard, WIlliam G., and Jonathan K. Pritchard, Mathematical models of running, *American Scientist* 82 (November-December 1994) 546–553.

The authors explain how to model human running speed very accurately, including correcting for wind. Their analysis leads them to conclude that the world-record time of Florence Griffith-Joyner in 1988 (10.49 seconds for the women's 100-meter sprint, with wind speed measured as 0.0 m/sec) was "anomalous," as it is likely that the run was in fact wind-aided to a considerable extent.

Solving the Quintic with Mathematica. Poster, 27" × 38". Wolfram Research, 100 Trade Center Dr., Champaign, IL 61820–7237; (800) 441–MATH. $2 plus postage.

Created for the International Congress in Zurich last summer, this attractive and detailed poster pictures the history behind several solutions to the quintic polynomial equation. Solutions to the quintic??? Yes, by Hermite, Kronecker, and Brioschi, in terms of elliptic modular functions, and by Klein in terms of hypergeometric functions. (The latter are already built into Mathematica, and the elliptic modular functions will be incorporated into the next release.)

Gerdes, Paulus, and Gildo Bulafo, *Sipatsi: Technology, Art and Geometry in Inhambane*, Instituto Superior Pedagógico, Maputo, Mozambique, 1994; 102 pp, (P). No ISBN number. U.S. distributor: Arthur B. Powell, Academic Foundations Dept., Rutgers University, Newark, NJ 07102; send check for $10 made out to "Arthur Powell (ISP Support)".

Sipatsi are woven basketry handbags decorated with ornamental bands, made in Inhambane Province of Mozambique. The first chapter of this book presents a brief ethnography, the second (taking up half the book) gives a catalog of strip patterns used to decorate *sipatsi*, and the third offers classroom ideas on mathematical exploration of *sipatsi* and patterns.

Freedman, David H., Lone wave, *Discover* (December 1994) 62–68.

Solitons, solitary waves that persist, result from a balance between dispersion and compression. Contraction of muscles may be explainable in terms of solitons, as they are "an efficient way to concentrate energy and get it to the right place." Perhaps a soliton also is responsible for prying apart the molecular strands of DNA; and perhaps gigantic soliton stars, rather than black holes, are at the center of galaxies that pour out enormous amounts of energy.

Faber, Scott, Jack's straws, *Discover* (December 1994) 81–83.

Jack Snoeyink (University of British Columbia) has devised sets of specially shaped pickup sticks that provide counterexamples to "folk" theorems. For one arrangement of six sticks, no stick can be removed (no twisting allowed) without disturbing the others. For a further array of 30 sticks, no stick, nor any pair of sticks, can be removed without disturbing the others (even with twisting allowed).

Arnott, Richard, and Kenneth Small, The economics of traffic congestion, *American Scientist* 82 ((september-OCtober 1994) 446–455.

Three separate perverse mathematical paradoxes illustrate that latent demand ("if you build it, they will fill it") and mispricing of congestion ("drivers do not pay for the time loss they impose on others") can thwart most attempts to relieve traffic congestion, whether by expanding capacity or diverting traffic to mass transit.

Mahoney, Michael Sean, *The Mathematical Career of Pierre de Fermat 1601–1665*, 2nd ed., Princeton University Press, 1994; xx + 432 pp, $18.95 (P). ISBN 0–691–0366–7.

The possibly imminent resolution of Fermat's Last Theorem has reawakened interest in Fermat himself. That interest is well served by this reissue in paperback of Mahoney's analysis of Fermat's work, with corrections and some updatings.

Schwartzman, Steven, *The Words of Mathematics: An Etymological Dictionary of Mathematical Terms Used in English*, MAA, 1994; vii + 261 pp, $27. ISBN 0–88385–511–9.

Here is a book from which everyone will learn something; it gives details about the etymological origins of mathematical terms. Systematically omitted are proper nouns derived from the names of people. An appendix groups together entries that are related by etymological root.

Guy, Richard K., and Robert E. Woodrow (eds.), *The Lighter Side of Mathematics: Proceedings of the Eugène Strens Memorial Conference on Recreational Mathematics and Its History*, MAA, 1994; viii + 367 pp, $38.50 (P). ISBN 0–88385–516–X.

This book commemorates a conference held in 1986, on the occasion of the University of Calgary's acquiring Eugène Strens's collection of 2,200 items about recreational mathematics. The editors have divided the papers into three sections: Tiling and Coloring, Games & Puzzles, and People & Pursuits. Many of the world's experts and aficionados in recreational mathematics are represented.

Friedman, Avner, and Walter Littman, *Industrial Mathematics: A Course in Solving Real-World Problems*, SIAM, 1994; xiii + 136 pp, $23 (P). ISBN 0–89871–324–2.

"Are calculus and 'post' calculus (such as differential equations) playing an important role in research and development in industry? Are these mathematical tools indispensable for improving ... automobiles, airplanes, televisions, and cameras? Do they play a role in understanding air pollution, predicting weather and stock market trends, and building better computers and communication systems? This book was written to convince you, by examples, that the answer to all the above questions is YES! Indeed, each chapter presents one important problem that arises in today's industry, and then gets down to studying the problem by mathematical analysis and computation." The chapters are adapted from the annual volumes of *Mathematics in Industrial Problems* edited by author Friedman; the topics are crystal precipitation, air quality modeling, electron beam lithography, development of color film, the catalytic converter, and the photocopier. The background required is multivariable calculus, but each chapter develops a little new mathematics; and exercises are included.

Devlin, Keith, *All the Math That's Fit to Print: Articles from the* Manchester Guardian, MAA, 1994; xvii + 330 pp, $29.50 (P). ISBN 0–88385–515–1.

From 1983 to 1989, author Devlin—who is the current editor of MAA's *Focus*—wrote a semimonthly column "Micromaths" about mathematics and computing for the English newspaper *The Manchester Guardian*. This book collects those columns. "I simply wrote about current events in mathematics, or what else interested me at the time." Good mathematical exposition is rare, and it is a treat to read these popular-press gems of 600 to 1,200 words. Cross-references and a topical index add to the value of the book.

Strang, Gilbert, Wavelets, *American Scientist* 82 (May-June 1994) 250–255.

This article is a too-brief trip through wavelets, with brief asides about the Fourier transform and the short-time Fourier transform. It also recounts the use of Fourier transforms in high-definition TV ("wavelets came too late to have a real chance") and why wavelets won over Fourier transforms in storing fingerprints.

Dunham, William, *The Mathematical Universe: An Alphabetical Journey through the Great Proofs, Problems, and Personalities*, Wiley, 1994; vi + 314 pp, $22.95. ISBN 0–471–53656–3.

This book, by the author of *Journey through Genius* (1990) (about the great theorems of mathematics), surveys of the author's choice of topics and anecdotes from mathematics. Your nonmathematical colleagues might enjoy its liveliness and lightness (it presumes only high-school algebra and geometry), and it may be suitable for a liberal-arts course in mathematics.

Schattschneider, Doris, Escher's metaphors, *Scientific American* (November 1994) 66–71.

Each generation must discover anew the pleasures and depths of M.C. Escher's prints, and their connections with mathematics. In this generously illustrated article, author Schattschneider concentrates on Escher's explorations of infinity.

van Leeuwen, J. (ed.), *Handbook of Theoretical Computer Science*, MIT Press, 1994. Vol. A: *Algorithms and Complexity*; ix + 996 pp, $50 (P). ISBN 0–262–22038–5 (hardcover), 0–262–72014–0 (P). Vol. B: *Formal Methods and Semantics*; xiv + 1273 pp, $60 (P). ISBN 0–262–22039–3 (hardcover), 0–262–72015–9 (P). The set: $90 (P). ISBN 0–262–22040–7 (hardcover), 0–262–72020–5 (P).

The division of topics between these two massive volumes reflects the division in theoretical computer science between "algorithm-oriented and description-oriented research." Vol. A treats models of computation, theories of complexity, data structures, and algorithms in many contexts. Vol. B is concerned with formal languages, automata, various rewriting systems, logic programming, proving programs, and models for databases, distributed computing,a nd concurrency. The 37 articles, by world experts, vary in their demands on the mathematical abilities of the reader; the articles are thorough but not easy. The volumes were first published in hardcover by Elsevier in 1990 and hence reflect knowledge as of then.

Schwartz, Richard H., *Mathematics and Global Survival*, 3rd ed., Ginn Press, 1993; xviii + 243 pp, (P). ISBN 0–536–58421–4.

This is the third edition of a student-motivating textbook on elementary descriptive statistics and probability; this edition adds student miniprojects and a chapter on the normal curve and confidence intervals. All examples and exercises are taken from global issues: arms, energy, health, hunger, pollution, population, poverty, resources, smoking, waste, and women's issues.

McLaughlin, William L., Resolving Zeno's paradoxes, *Scientific American* (November 1994) 84–89.

Using technical results from the theory of nonstandard analysis, the author refutes Zeno's paradoxes and constructs an alternative theory of motion.

Horgan, John, Global politics: Mathematicians collide over a claim about packing spheres, *Scientific American* (December 1994) 32, 34.

Controversy continues over whether Wu-Yi Hsiang (UC—Berkeley) has proved the Kepler conjecture about optimal sphere-packing. Most experts think not. "It is harder to check proofs than it used to be," says Chandler Davis (University of Toronto).

NEWS AND LETTERS

LETTERS TO THE EDITOR

Dear Editor:

Concerning my paper, "In the Gaussian Integers, $\alpha^4 + \beta^4 \neq \gamma^4$" (April 1993), there is an error in the ninth line from the top. The sentence should read as follows:

Now $N(a^2+b^2) \leq N(a^2)+2\,N(b^2) + 2\,N(a^2)\,N(b^2)$, since for positive integers m and n, not both 1, $m+n < 2\,m\,n$.

I am indebted to William P. Wardlaw, of the U.S. Naval Academy, for reading the paper carefully enough to find this hole in the proof. Please allow me to offer my apologies.

James T. Cross
University of the South
Sewanee, TN 37383

Dear Editor:

Regarding John O. Kiltinen's article, "How Few Transpositions Suffice? ...You Already Know!" (February 1994), here is a shorter proof that the standard expression is minimal. For $\alpha = \sigma_1 \circ \cdots \circ \sigma_r \in S_n$ where the σ_i are disjoint cycles of respective lengths k_i (*including* the trivial cycles of length 1, if any), define the index of α to be $\Sigma_i(k_i-1) = n-r$. (This is expression (2) in Kiltinen's paper. It's called the *index* in, e.g., Mostow-Sampson-Meyer, *Fundamental Structures of Algebra*, McGraw-Hill, 1963.) Next, if π is any permutation and $\tau = (a,b)$ is any transposition, then a,b are either (1) in the same cycle σ of π or (2) in different cycles of π. In case (1), say $\sigma = (a,x_1,\ldots,x_s,b,y_1,\ldots,y_t)$, then $\tau\sigma = (a,x_1,\ldots,x_s)(b,y_1,\ldots,y_t)$, hence $\tau\pi$ has one more cycle than π. In case (2), a similar argument shows that $\tau\pi$ has one less cycle than π. In either case, $\mathrm{ind}(\tau\pi) = \mathrm{ind}(\pi) \pm 1$. Since the identity permutation has index 0, it follows by induction that any product of m transpositions has index $\leq$ m; hence the standard expression is minimal.

The equation $\mathrm{ind}(\tau\pi) = \mathrm{ind}(\pi) \pm 1$ further implies by induction that any product of m transpositions has index $\equiv$ m (mod 2), which in turn implies the parity theorem for S_n. This method is used by Jacobson in his *Lectures in Abstract Algebra*, Vol. I, Van Nostrand Reinhold Co. Inc., New York, 1951, p. 36.

David M. Bloom
Brooklyn College of
CUNY
Brooklyn, NY 11210

Professor Bloom's proof is indeed shorter and more elegant. It conforms to the mathematical aesthetic that prizes brevity and elegance above all else. However, from a pedagogical point of view, I am not so sure that briefest is always best, especially when students are first working with a new idea. Jacobson's very succinct proof of the parity theorem that Professor Bloom cites was likely known to some of the half-dozen writers of undergraduate texts that I cited, even though I missed it. Nevertheless, they opted for the longer, transposition-pushing proof, and I think with good pedagogical reason. The single, deft, studied tap of the diamond cutter's hammer is elegant to behold as it releases its gems, but there is also something satisfying--and instructive--in getting one's hands, like a potter's, into the wet clay from which we turn out our everyday mathematical crockery. My note shows that a well-known potter's wheel can be used to spin out not only bowls, but also vases.
--John O. Kiltinen

Dear Editor:

In a recent Note, "Using Complex Solutions to Aid in Graphing" (April 1993), C. Bandy used the complex roots

$\alpha \pm \beta i$ to graph the quadratic $y = ax^2 + bx + c$. Assume that $a > 0$.

Measuring α is not difficult. Since

$$y(x) = ax^2 + bx + c = a(x - \alpha + \beta i) = a((x-\alpha)^2 + \beta^2) \geq a\beta^2$$

y has its minimum at $x = \alpha$ (for $a > 0$). So, α is the x-coordinate of the minimum point or vertex.

Other intriguing questions are: Can the imaginary part be measured? and, If it can be, how can it be measured? One way to see that the answer to the first question is YES is the following simple construction displayed in the figure. From the point $(\alpha, 0)$, which lies directly below the minimum, draw a vertical line to $(\alpha, 2D)$ where D is the y-coordinate of the minimum. Then draw a line parallel to the x-axis at $y = 2D$ to meet the curve, and drop a perpendicular line from the meeting point to the x-axis. The x-coordinate there is $\alpha + \beta$. So, β is the distance from $x = \alpha$ to $x = \alpha + \beta$.

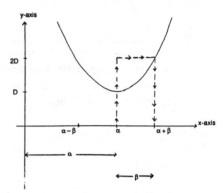

Construction for imaginary part β.

David L. Farnsworth
Rochester Institute of
Technology
Rochester, NY 14623